5 SCIENTISTS, 7 ENGINEERS,
AND 2 AUTHORS MAKING
YOUR SCIENCE FICTION
DREAMS COME TRUE

5 SCIENTISTS, 7 ENGINEERS, AND 2 AUTHORS MAKING YOUR SCIENCE FICTION DREAMS COME TRUE

LAUREN YUN

NEW DEGREE PRESS

5 SCIENTISTS, 7 ENGINEERS, AND 2 AUTHORS MAKING YOUR
SCIENCE FICTION DREAMS COME TRUE

ISBN   978-1-64137-192-6   *Paperback*

        978-1-64137-193-3   *Ebook*

CONTENTS

INTRODUCTION

"Come with me
And you'll be
In a world of
Pure imagination."

—*WILLY WONKA, WILLY WONKA AND THE*

CHOCOLATE FACTORY (1971)[1]

I love movies.

But growing up, I hated going to sleepovers for one main reason:

---

[1] "Pure Imagination [From Willy Wonka And The Chocolate Factory] Lyrics," *Lyrics.Com*, accessed January 15, 2019, https://www.lyrics.com/lyric/27628631.

I always <u>hated</u> the movies people chose to watch.

I never liked comedy movies because I never thought they were intellectually stimulating enough, and I never enjoyed the cheesy romance movies where the last scene of the film is a guy and a girl kissing in the rain. I am an action-adventure movie person. I love fast plots with fight scenes and people uniting against a common threat. I love *Star Wars*, *Iron Man*, *Jumanji* (the new and old one), *Star Trek*, *The Matrix*, and more.

By the time I was in middle school, I started to notice a recurring theme in my movies—they were all science fiction.

I always loved the futuristic element about these movies. Even though they take place in different times, they all have concepts that seem unimaginable.

As a child, I was fascinated with the future: more advanced technology, holograms you could touch, cell phones, and teleportation. My love for technology stems from the idea that I always want to advance. I am always looking for the next step. I am young and have grown up in a world of many science-fiction movies, especially, ones dealing with the future. *Ready Player One* seems unlikely to some, but it actually contains many elements that are possible today or will be in the near future.

In fact, science fiction has given me a passion for a career in science.

Because I have grown up with technology, I'm hoping to connect the "futuristic" elements of science fiction movies to the world today.

* * *

"But that's not real science."

When I share with people that I became interested in STEM through science fiction films, I'm often told that the films I like are more fiction than science or more fantasy than science fiction.

And admittedly, it's a difficult distinction to argue.

Is light speed real?

Is a light saber actually something we could one day fight with?

Could you have a robotic hand that functioned as well or better than a real human hand?

I want to believe they are all real possibilities, but I don't know.

So I decided to write this book.

* * *

In seventh grade, I had the nickname "The Apple Stalker."

It was a running joke with the kids on my bus on the way home from school. When I tell people this story, they usually think I'm obsessed with fruit; however, the nickname stems from the fact that I have always been obsessed with Apple's technology. People always knew they could come to me to ask about what new features would be on the next iPhone, what the colors would be or when Apple's next showcase presentation would be. I am lucky enough to say in seventh grade I already had an iPhone and laptop, and I brought them everywhere with me. Surprisingly, most of my friends don't considered me "addicted" to technology, but I would always have these devices because they made my life so much easier and interesting. I always used my phone to take hilarious videos and photos of my friends.

Thankfully, I have grown out of my video blogging and photography stage (I was awful at it), but I am glad people still think of me as someone they can seek out to learn more about technology.

For me, part of this book is about ensuring that we don't become so disillusioned by the naysayers that we stop believing these technologies in science fiction movies and books have the potential to become real.

I used to believe that AI (Artificial Intelligence), voice technology, and holographic images were still extremely futuristic. I didn't even realize how much of this is incorporated into my life today. Siri is AI. In March 2017, seven hundred million people were using iPhones.[2] That means at least seven hundred million people have access to Siri, who is much more capable than you may think of helping people.

If you're lucky enough to be a fully healthy human like me, Siri does not seem like a huge deal. But for others, Siri is revolutionary. People who are not able to type, such as individuals with mental or physical disabilities, can now communicate with millions of people around the globe.

Technology like Siri, Amazon's Alexa, and Google Homes are stepping stones to even more advanced AI that could help human life like no other. Did you know Siri can call you an Uber? Or check the number of calories in a Big Mac? Did you know Alexa can help you meditate and work on your mental state? Or quiz you on Marvel movie trivia? Did you know Google Home can find your lost phone? Or find you

---

2   Reisinger Don, "Here's how Many iPhones are Currently Being Used Worldwide," *Fortune*, accessed January 10, 2019, http://fortune.com/2017/03/06/apple-iphone-use-worldwide/.

a plumber? People don't realize that AI, voice technology, and technology assisting humans are NOW. We are in what some may believe is the future. Google's assistant runs on four hundred million devices.[3]

Some of my friends have an Amazon Alexa for every room of their house. The reality is, we are becoming more AI and voice driven. I think discussing this kind of technology, especially Artificial Intelligence, makes people think of robots that can function exactly like human beings. A large misconception is that AI can do everything a human being can do. For some AI researchers, that is the ultimate goal, but we have devices that can do an insane amount for us.

If you'd seen the movie *2001: A Space Odyssey* and thought "that's just fantasy," maybe you never would have thought to try and create the technology that is in our phones and in our homes today.

Being curious and inspired is imperative for the next generation.

* * *

---

3   Lardinois Frederic, "Google says it sold a Google Home device every second since October 19," *Tech Crunch*, accessed January 10, 2019, https://techcrunch.com/2018/01/05/google-says-it-sold-a-google-home-device-every-second-since-october-19/.

I am here to tell you that all this "futuristic" technology is actually much more accessible to you today than most of us may realize, and people are working to improve it to help human beings. Many of the people I interviewed and researched for this book believe AI and other technologies will only improve to better human life. And it already has.

Yes, I remember flip phones, Blackberries, and life without constant technology, but I am one of the last people to remember that. Technology has become so heavily integrated into my life sometimes I even forget that having a laptop was not a societal norm thirty years ago. Futuristic technologies seen in movies, comic books, and other forms of pop culture have made it seem like light speed is only possible "a long time ago in a galaxy far, far away," and I am here to tell you that is not true.

It is possible.

It's all possible, and technologies you thought were lightyears away are here today.

* * *

You do not have to read this book in order. You can jump to topics that resonate with you and come back to others later. The following chapters contain stories from people I have spoken with ranging from entrepreneurs to scientists, all of whom are leading figures in their industry.

We'll discuss concepts ranging from Siri to light sabers, and I hope you find one you are interested in learning about its progression. People who love sci-fi movies, the future, and technology will hopefully find a subject or an aspect of their favorite film that is being developed in the real world today.

This book is for sci-fi nerds or really anyone who just wants to discover current and upcoming developments in science and technology.

So flip through, find something that excites you, and learn more about how some of your favorite sci-fi movie concepts are coming true and how futuristic technologies are actually making this world a better place. You will discover that concepts you may have thought were unimaginable are actually impacting real people right now.

Come with me, and we'll uncover a world of pure futuristic possibilities that are on the horizon and will let us live in a world of science fiction.

Defining Science, Fiction & Fantasy

WHAT DEFINES SCIENCE FICTION?

You may have difficulty with this question—possibly because you've never really thought about it. Ask yourself this: Is *Star Wars* science fiction? Of course the answer is yes ... right? There are spaceships, aliens, laser blasters and light sabers. Without a doubt, *Star Wars* is science fiction. Well, not entirely. So, what makes *Star Wars* not science fiction?

Science Fiction. Two words. Let's look at them separately.

SCIENCE:

Science is defined in Webster's New World Dictionary as, "systemized knowledge derived from observation, study, etc." This is to say that science knows things are real because they are observable in the real world around us. Not only must they be observable, but they must be repeatable. A fluke or random event isn't considered a scientific fact unless it can be repeated, simulated and observed by trained scientists whose scrutinizing gaze can confirm that it does, in fact, exist.

FICTION:

Now let's look at the word fiction in the same dictionary. Fiction is defined as, "any literary work with imaginary characters and events." Certainly, *Star Wars* fits into this category: it's fictitious. That's an undeniable fact. But science fiction is fiction (imaginary characters and events) bound by the observable and repeatable laws of science. Granted, the laws of science are always expanding and changing, and screenwriters should feel free to develop stories containing science that does not yet, or may never, exist.

Now look at the major theme of the *Star Wars* universe: The Force.

What is it? How does it work? We're told it's in us, all around us—a magical force that enables certain characters, Jedi Knights, to perform such amazing feats as: levitation, telekinesis, mind control and lightning bolts from finger tips. The list goes on and on, with no scientific explanation in sight. The Force alone separates *Star Wars* from the world of pure science fiction. This doesn't make *Star Wars* bad science fiction; it simply makes it a mixture of sci-fi and fantasy.

FANTASY:

I came across this definition of fantasy in a Google search: "the faculty or activity of imagining things, especially things that are impossible or improbable." Fantasy comes into play with a lot of sci-fi movies. In science fiction, your creations need to make sense within the natural laws of the universe (our universe). Going back to Superman, how does he fly? Does he have a jet pack? Wings? Does his body fill with helium? No, he simply wills himself to fly, and he does so because Earth's yellow sun gives him super powers (Krypton's sun was red). This is pure fantasy. There is no scientific reason for Superman's powers, thus he'll remain an icon of the fantasy genre.

Science fiction and fantasy are becoming reality. Although their definitions imply they are unrealistic or not possible, scientists and innovators are making the impossible possible. I hope this book takes you along a journey where you will learn some of your sci-fi dreams could come true.

PART 1

GALAXIES (NOT SO) FAR FAR AWAY: TECH ON THE SCREEN AND THE PAGE

CHAPTER 1

COULD WE REALLY JUMP TO LIGHT OR WARP SPEED?

———

[The Millennium Falcon is speeding away from Tatooine, being chased by Imperial cruisers.]

Han Solo: Stay sharp! There are two more coming in. They're going to try to cut us off.

Luke Skywalker: Why don't you outrun them? I thought you said this thing was fast!

Han Solo: Watch your mouth, kid, or you'll find yourself floating home! We'll be safe once we make the jump to hyperspace.

Besides, I know a few maneuvers. We'll lose them. This is where the fun begins.

Ben Kenobi: *How long before you make the jump to light speed?*

Han Solo: *It'll take a few moments to get the coordinates from the navicomputer.*

Luke Skywalker: *[frantic] Are you kidding? At the rate they're gaining—*

Han Solo: *Traveling through hyperspace ain't like dusting crops, boy! Without precise calculations we could fly right through a star or bounce too close to a supernova and that'd end your trip real quick, wouldn't it?*

Luke Skywalker: *[points to an alarm on the control panel] What's that flashing?*

Han Solo: *We're losing our deflector shield! Go strap yourself in, I'm going to make the jump to light speed.*

—*STAR WARS EPISODE IV: A NEW HOPE* (1977)[4]

---

4 "View Quote ... Star Wars Episode IV: A New Hope ... Movie Quotes Database," *Moviequotedb.com*, accessed January 13, 2019, http://www.moviequotedb.com/movies/star-wars-episode-iv-a-new-hope/quote_29884.html.

Captain James Kirk: *Scotty, I need warp speed in three minutes or we're all dead!*

— STAR TREK: THE WRATH OF KHAN (1982)[5]

Who doesn't want to "jump to light speed" or "warp speed" and zoom to another galaxy in minutes? It's part of the magic of science fiction space movies—connection to long lost planets and people. And frankly, if travel by light speed isn't possible, it makes the boring decades' long travel required, instead, seem way less interesting crammed into a two hour movie.

When you begin doing a little research online into the potential of humans traveling at light speed or warp speed, arm chair physicists might give you a somewhat disappointing answer.

But not Geraint Lewis.

I found Lewis through a lecture he gave on how light speed/ warp speed could be a reality—one of my *Star Wars* dreams coming true.

---

5 "Scotty, I Need Warp Speed In Three Minutes Or We're All Dead! - Star Trek: The Wrath Of Khan," *QuoteGeek*, accessed January 15, 2019, http://quotegeek.com/quotes-from-movies/star-trek-the-wrath-of-khan/840/.

If you ask some experts, the ability to travel from one planet to another in minutes is apparently impossible; thankfully Lewis does not believe this to be true.

"Everyone knows that relativity has a speed limit, the speed of light. That essentially comes from special relativity. But when you have general relativity, the question of 'What does that speed limit mean?' is slightly more complicated. It's a local speed limit, not a global speed limit. So, theoretically, you can travel at any speed."

Lewis does some intricate and fascinating work studying a level of physics few will ever understand—me included. His desire to become a scientist started at a very young age and continued through his adult life.

Thankfully, he was willing to simplify things for me.

"The big issue is that you need to warp and bend space and time in a particular way. You need this exotic material, like dark energy. It has a negative pressure, and we do not know any devices that could hold and shape the energy to give you a warp drive. But theoretically, on paper, it is possible."

That's promising. But I wondered why so many others out there seem to say it's impossible.

He thought for a second: "Well, whether it is actually physically possible is a different question."

* * *

Unfortunately, Lewis was unable to offer me anything more certain than that. Turned out proving it was physically possible was even above his pay grade.

While I really *hope* it is physically possible, there is no research to say so… yet.

So, what does that mean for future space travelers like me? Do we have a shot at light speed travel and is it possible for me to complete my *Star Wars* dream?

Theoretically Possible? Yes.
Physically Possible? Unclear.
Time Horizon? Unknown

Han Solo: Punch it, Chewie.

—*STAR WARS EPISODE V: THE EMPIRE STRIKES BACK* (1980)[6]

---

6   "View Quote … Star Wars Episode V: The Empire Strikes Back… Movie Quotes Database," *Moviequotedb.com*, accessed January 13, 2019, http://www.moviequotedb.com/movies/star-wars-episode-v-the-empire-strikes-back/quote_22681.html.

CHAPTER 2

IS IT POSSIBLE TO HAVE A REAL HOLOGRAM?

Princess Leia: "*General Kenobi. Years ago you served my father in the Clone Wars. Now he begs you to help him in his struggle against the Empire. I regret that I am unable to convey my father's request to you in person, but my ship has fallen under attack, and I'm afraid my mission to bring you to Alderaan has failed. I have placed information vital to the survival of the Rebellion into the memory systems of this R2 unit. My father will know how to retrieve it. You must see this droid safely delivered to him on Alderaan. This is our* *most desperate hour.* **Help me, Obi-Wan Kenobi. You're my only hope.**"

—*STAR WARS EPISODE IV: A NEW HOPE* (1977)[7]

The idea of these hologram visuals has found its way into multiple movies from the *Star Wars* films to *Iron Man*. What are they and could they be real?

It turns out the visualization of Leia, which ultimately leads her to find her brother Luke, is a 3-D image, not a hologram.

I learned this from Curtis Broadbent, a professor at the University of Rochester.

Broadbent attended Brigham Young University, and a professor there suggested he to apply to Rochester. He wasn't even originally working on 3-D images. Instead he was studying atomic physics.

"The lab I was in isn't a display lab. We were doing atomic physics, studying what happens when you send certain frequencies of light into an atom and how to get atoms to behave in special ways."

---

7   "View Quote ... Star Wars Episode IV: A New Hope ... Movie Quotes Database."

Broadbent has been helping make a volumetric 3-D image a reality. Although it will not be in the near future, Broadbent's work could lead to 3-D images (known to the general public as holograms) coming out of phones, laptops, and tables at home.

"One of the things we had been studying was the two photon absorption process. If you send in two frequencies of light that are perfectly tuned, you can cause the atom to go from a ground state to an intermediate level to an upper excited level. And an atom in an upper excited level can decay through multiple decay paths, one of which emits blue light.

"So if you take two infrared lasers, you can excite these atoms up a level where they will emit visible light in a localized way. We were working with this process for a long time and then it occurred to me—this process can be used to create a volumetric display! We found out later that other people had similar ideas in the eighties, but we added some new techniques to enhance the process. It was a tool we were already using, and we simply realized could be applied to a display application."

* * *

His inspiration came from his desire to visualize shapes he learned about in his classes.

"The original inspiration actually came when I was in high school. I took three-dimensional geometry in my junior year. I remember thinking 'Wow, it would be so neat to visualize these shapes,' in real three dimensions so you could walk around these three-dimensional shapes. I just thought that would be really cool. I didn't really know how to do that. I was kind of thinking of that old idea when I had the inspiration to cross the laser beams and use two-photon absorption induced fluorescence. That's what really inspired us. Maybe the invention lends itself to *Star Wars* or *Iron Man* or all those other entertainment references, but the original inspiration came from a desire to visualize simple mathematical shapes. There is beauty in simplicity—and to see it in real space is mesmerizing."

Personally, I was excited to hear that Broadbent's original inspiration came from a class he took in high school. He gives me hope that my ideas now could turn into something that could change the world in the future.

"While our technology only works in a vacuum, I think the real game changer will be when someone figures out how to create a high-resolution volumetric image in free space, so you could potentially put your hand in there. A few technologies currently can create volumetric images in real space, but some are low resolution and others just aren't safe. One technology uses super short-pulse lasers to ionize the air, and then when the air is ionized it is fluorescent and usually emits white light only, but those lasers are very expensive. They are also very dangerous. You certainly would not want to put your eye anywhere near where those lasers. I tend to think that is not the route forward, but I have heard they are working on color. So, if you're not going to do it that way, you have to come up with some other approach, but I do think it will be a really big thing—a free space volumetric display that's safe for all viewers."

From *Iron Man* to *Star Wars*, the 3-D volumetric display that will likely come in the future.

Broadbent is on the way to making that happen; however, he says it will still take some time. There are road blocks to overcome, but Broadbent knows what scientists will need to develop next to make this a reality.

"It will still take a while to develop real space volumetric displays. There are many challenges to overcome. Our current design requires use of a vacuum chamber, so that will not be part of a phone anytime soon."

* * *

As Broadbent continues to work on his display, I have hope that I will be able to see Princess Leia saying her iconic message soon in real space. His work is fascinating, and people have already seen so many similar concepts in movies.

The potential applications are endless and could easily change the advertising industry, education, and many other aspects of the world.

He and his team are working hard to make many *Star Wars* dreams come true.

Theoretically Possible? Yes.
Physically Possible? Yes.
Time Horizon? A while.

Pepper Potts: What is all of this?

Tony Stark: This is, uh...

[Different profiles appears in holographic form floating in the air in front of Stark and Pepper.]

Tony Stark: This.

[Screens appear of Captain America in action, the Hulk roaring as he attacks the Army at Culver University. Another is of Loki and the Tesseract, to which Stark and Pepper look on in awe.]

Pepper Potts: *I'm going to take the jet to DC tonight.*

Tony Stark: *Tomorrow.*

Pepper Potts: *You've got homework. You've got a lot of homework.*

—THE AVENGERS (2012)[8]

---

8   "The Avengers (2012)," *Imbd*, accessed January 13, 2019, https://www.imdb.com/title/tt0848228/characters/nm0000569.

CHAPTER 3

IS A REAL LIGHTSABER POSSIBLE?

———

Luke Skywalker: No, my father didn't fight in the wars. He was a navigator on a spice freighter.

Ben Kenobi: That's what your uncle told you. He didn't hold with your father's ideals. Thought he should have stayed here and not gotten involved.

Luke Skywalker: You fought in the Clone Wars?

Ben Kenobi: Yes, I was once a Jedi Knight the same as your father.

Luke Skywalker: I wish I'd known him.

Ben Kenobi: *He was the best starpilot in the galaxy, and a cunning warrior. I understand you've become quite a good pilot yourself. And he was a good friend. Which reminds me, I have something here for you. You father wanted you to have this when you were old enough, but your uncle wouldn't allow it. He feared you might follow old Obi-Wan on some damn fool idealistic crusade like your father did.*

C-3PO: *Sir, if you'll not be needing me, I'll close down for awhile.*

Luke Skywalker: *Sure, go ahead.*

[C-3PO shuts down.]

Luke Skywalker: *[to Ben] What is it?*

Ben Kenobi: *Your father's lightsaber. [He turns it on and hands it to Luke, who begins swinging it around while Ben continues.] This is the weapon of a Jedi Knight. Not as clumsy or random as a blaster. An elegant weapon, for a more civilized age. For over a thousand generations, the Jedi Knights were guardians of peace and justice in the Old Republic. Before the dark times. Before the Empire.*

Luke Skywalker: *[turns off the lightsaber] How did my father die?*

Ben Kenobi: A young Jedi named Darth Vader, who was a pupil of mine until he turned to evil, helped the Empire hunt down and destroy the Jedi Knights. He betrayed and murdered your father. Now the Jedi are all but extinct. Vader was seduced by the dark side of the Force.

Luke Skywalker: The Force?

Ben Kenobi: The Force is what gives a Jedi his powers. It's an energy field created by all living things. It surrounds us, penetrates us. It binds the galaxy together. [to R2-D2] Now, let's see if we can't figure out who you are, my little friend. And where you come from.

—STAR WARS EPISODE IV: A NEW HOPE (1977)[9]

Let's face it. The Force will never be real. However, other aspects of the *Star Wars* universe could become a reality.

Although the Princess Leia message may be iconic, the lightsaber is the ultimate *Star Wars* symbol.

Don Lincoln researched how the lightsaber could become real (and how you could slay your enemies with it).

---

9   "View Quote ... Star Wars Episode IV: A New Hope ... Movie Quotes Database."

From Kylo Ren's newest model, shaped like a cross, all the way back to the iconic Skywalker lightsaber that Anakin, Luke, Rey, and Finn have all used, the light saber is a *Star Wars* icon. Most of the lightsabers you see in the real world are just toys for kids or adults and cannot actually be used to hurt anyone. But it would be pretty amazing to see how a real lightsaber could work. Right? You know, just in case you need to battle the next Darth Vader.

* * *

What is a lightsaber? Well Don Lincoln, a senior scientist at the US Department of Energy's Fermilab, is doing research on particle physics for the Large Hadron Collider. He also writes about science for the public. He wrote an article for space.com in 2015 about new hopes of building a real lightsaber.[10]

"Given the name and appearances, the first obvious thought is that perhaps lightsabers consist of some kind of laser. However, this hypothesis is easy to rule out. Lasers don't have a fixed length, as you can determine using a simple laser pointer. Further, unless the light is somehow scattered, a laser

---

10 Lincoln Don, "Is a Real Lightsaber possible? Science Offers a New Hope," *Space.com*, accessed January 10, 2019, https://www.space.com/31361-building-a-real-lightsaber.html

is essentially invisible as it passes through the air. Neither of these characteristics describes a lightsaber."[11]

Although a laser is not the answer, maybe plasma is. Plasma is another state of matter outside the common three: solid, liquid, and gas.

"A more realistic technology is a plasma. Such a material is created by stripping a gas's atoms of their electrons, a process called ionization. This stripping causes the material to glow. A plasma is a fourth state of matter, after the familiar three states of solid, liquid and gas. You have seen examples of plasmas all of your life. The glow of a fluorescent light is a plasma, as are neon lights. Some plasmas can actually generate considerable heat. These are called plasma torches. The principle is the same as a lightbulb, but with more electrical current involved. There are many ways to make a plasma torch, but the simplest one employs two electrodes and a flowing material, usually a gas such as oxygen, nitrogen or something similar. A high voltage on the electrodes ionizes the gas, converting it into a plasma.[12]

"Are lightsabers simply ultrahot plasma tubes, then? Not necessarily, as a plasma acts somewhat like a hot gas, which

---

11 "Is a Real Lightsaber possible?"
12 Ibid.

expands and cools, just like an ordinary fire (which is often a plasma, albeit an incomplete one, as can be seen by the fact that it glows). So if a plasma is the base technology of a light saber, it needs to be contained.[13]

"Luckily there is a mechanism for doing this. Plasmas, being composed of charged particles (some with very high velocities), can be manipulated by magnetic fields. In fact, some of the more promising technologies involved with nuclear fusion research use magnetic fields to contain plasmas. The temperatures and total energy contained in fusion plasmas are so high that they would melt their metal containment vessels. So this is promising for lightsabers, too. Strong magnetic fields, coupled with a very hot and dense plasma provide a candidate method for creating a lightsaber."[14]

Lincoln's description of a realistic light saber quickly reminded me of a scene in *The Force Awakens*. The battle between Rey and Kylo Ren in the snow shows how hot the light saber really is. Every time a saber touches the snow, the audience can hear the snow vaporizing with a sizzling sound.

"However, we're not done. If we had two magnetically contained tubes of plasma, they'd pass right through one another

---

13   Ibid.
14   Ibid.

... so no epic lightsaber duels. For that, we need to figure out a way to make a solid core for the sabers. And the material that makes up the core would have to be impervious to the hot temperatures."[15]

So yes, we wouldn't be able to fight with these lightsabers, but still having a real looking one would be a dream come true for most *Star Wars* fans. And probably ideal since people couldn't actually duel with them.

This idea has not been commercialized in any way; it is very complicated, but if someone built a prototype, it would still make some of my childhood dreams come true.

* * *

However, in 2017, Lincoln wrote an article for CNN on the science behind the lightsaber titled "Unfortunately, the Force may not be with you."[16] Lincoln discusses some of the issues with creating this kind of device.

"From *The Phantom Menace*, we can deduce that the Jedi's lightsaber must wield at least 20 megawatts of power, which

---

15 Ibid.

16 Lincoln Don, "Unfortunately the Force May Not Be With You," *CNN*, accessed January 10, 2019, https://www.cnn.com/2017/12/17/opinions/ star-wars-science-possibilities-opinion-lincoln/index.html

is the energy needed to run approximately 14,000 American households, all stored in a device small enough to be held in the human hand. There are no known energy sources with that kind of capability, except perhaps antimatter. Even nuclear power won't work." [17]

Although I still have my hopes up, it seems unlikely in the near future that we will be able to put the power of 14,000 households into one person's hand. Lincoln goes on to explain that in addition to the energy problem, there are even more difficulties with building a real world lightsaber.

"By ignoring this energy problem, various technologies have been suggested for how a lightsaber might work, but most center on the possibility that they are plasmas with perhaps a ceramic core. Then there is the issue of why the immense heat of these plasmas do not char the hands of the Jedis. All in all, the lightsabers and the possible technical solutions raise more questions than they solve. Reality check: Sorry, Luke, Han was right when he said 'ancient weapons are no match for a good blaster.'"[18]

I am thankful that Lincoln wrote about this to educate fans like me; however, it was a little heartbreaking to realize this

---

17 "Unfortunately the Force May Not Be With You."
18 Ibid.

may not be a reality in my lifetime or even ever. Something that has always been interesting to me about *Star Wars* is that it takes place a long time ago, but everything seems futuristic from lightsabers to light speed. Nothing in these movies seems old.

Star Wars takes people in the theaters to another world, and I guess that's the magic of it; it doesn't matter if it's the future or the past. It's just a different and extraordinary place where lightsabers are real and people can travel from one planet to another in minutes. Even though lightsabers aren't real, every time I watch a *Star Wars* movie I believe they are.

Theoretically Possible? Maybe.
Physically Possible? No.
Time Horizon? Never.

C-3PO: Sir, the possibility of successfully navigating an asteroid field is approximately 3,720 to 1.

Han Solo: Never tell me the odds.

—STAR WARS EPISODE IV: A NEW HOPE (1977)[19]

---

19 "View Quote ... Star Wars Episode IV: A New Hope ... Movie Quotes Database."

PART 2

THE FUTURE OF HEALTH

CHAPTER 4

THE FORCE BEHIND CREATING A REAL "SKYWALKER HAND"

―――――

[Darth Vader cuts off Luke's hand. Luke attempts to escape by climbing out onto a projection in the middle of a long ventilation shaft.]

Darth Vader: *There is no escape! Don't make me destroy you. Luke, you do not yet realize your importance. You've only begun to discover your power! Join me, and I will complete your training! With our combined strength, we can end this destructive conflict and bring order to the galaxy.*

Luke Skywalker: *[angrily] I'll never join you!*

Darth Vader: *If only you knew the power of the Dark Side! Obi-Wan never told you what happened to your father.*

Luke Skywalker: *He told me enough! He told me you killed him!*

Darth Vader: *No. I am your father!*

Luke Skywalker: *[shocked] No... No! That's not true! That's impossible!*

Darth Vader: *Search your feelings. You know it to be true!*

Luke Skywalker: *[overwhelmed, crying]* **NOOOOOOOOOOOOOOOOOOOOOO! NOOOO!**

Darth Vader: *Luke, you can destroy the emperor. He has foreseen this. It is your destiny! Join me, and together, we can rule the galaxy as father and son! Come with me. It is the only way.*
—*STAR WARS EPISODE V: THE EMPIRE STRIKES BACK* (1980)[20]

---

20 "View Quote ... Star Wars Episode V: The Empire Strikes Back... Movie Quotes Database."

Two out of the three *Star Wars* trilogies has a Skywalker lose a hand in the second movie. Luke loses his hand after his father, Darth Vader, chops it off in their iconic showdown, and Anakin loses his hand to Count Dooku in *Attack of the Clones*.

Both Luke and Anakin's prosthetics are so futuristic that they both are able to use their hands with the same mobility after their battles. Luke's hand looks like a real hand with skin, and he has full use of all his fingers. Anakin's is metal and looks more robotic.

However, this is not the reality for prosthetics and amputees today. We are trying to make significant progress toward a world much like a galaxy far, far away where multiple generations of Skywalkers can benefit from advanced work in prosthetic hands.

* * *

Gil Weinberg, a professor at Georgia Institute of Technology, combined his love of engineering and music to create a hand for amputee Jason Barnes that allows him to play the piano again.

Weinberg's cross between engineering and music started from his childhood.

"Originally, I am from Israel, and then I did my graduate school at MIT. I was a musician first before I became a computer scientist, and I was a musician that combined computer science for a long while before I got into robotics. So that was my evolution of interests. I played piano. Even as a teenager, I was interested to see what happened when I wrote a computer program that responded to me like no human could. I was always looking to see if I could be inspired or if I could think about music differently, play music differently, if there was some process humans don't use. I wanted to write computer software that would respond to the piano.

"When I came to Georgia Tech, I was missing the acoustic sound. The one thing that could combine AI and algorithm, enterprise, and exploration of what notes are being played with acoustic sound and visual cues, physical space and visual cues and embodiment and enrichment of actually playing with an entity was obviously robots, and that's why I started to work with robots that listen like humans and play like machines. They understand music like humans do."

* * *

In 2012, Barnes was severely shocked during a work accident. Doctors had to amputate his arm below his elbow. Barnes set out to find someone who could help him play music again.

When Barnes found Weinberg, Weinberg set out to help Barnes gain more mobility in his hand again.

"The trajectory was music, computers, robots, and only then prosthetics. Jason Barnes lost his arm. He was devastated. He looked for people who could build robotics with music because he wanted a musical arm. But then he found out how I take some of my ideas and algorithms from separate entities. A couple of years ago we actually pushed him even further. With the Skywalker [that's what they call Barnes's upgraded prosthetic], we used machine learning to learn Jason's intentions from the muscles, so it's much more accurate.

"We are trying to commercialize it, so hopefully very soon people will be able to buy it—not only amputees but hopefully everyone. We've actually seen an interesting trend where people are starting to see that these things can grow and be relevant for everyone. We are now trying to commercialize it for all kinds of things. I believe we could have a product on the market within two years, but it'll be expensive, so not everyone can get it."

When I heard about this, I was completely blown away by the work. Barnes won the lottery here. He found the exact professor he needed to continue to play the music he loved. Through stories I heard of amputees and even in movies, I thought if you lost a body part like a hand or a foot, many activities that took a lot of mobility, such as playing piano, were over.

In countless surfing and war movies, people have to learn to adapt to their body without certain limbs, or they never fully recover at all. Barnes was also very surprised and thankful that Weinberg gave him so many abilities that normal prosthetics do not have.

"It's completely mind-blowing. This new arm allows me to do whatever grip I want, on the fly, without changing modes or pressing a button. I never thought we'd be able to do this."

* * *

Weinberg and Barnes decided to continue their work together, even after the hand. In a video I watched about their work together, they discussed a prosthetic arm that Barnes uses to play the drums again. Barnes plays with his real hand and the prosthetic hand.

"This came to be after Jason Barnes became very successful and well-known. We traveled around the world. We had thousands of performances on four continents, and then we thought 'why only amputees, why only Jason Barnes?' If the arm can play like no human can with its mechanical abilities, it can play faster than any human. Barnes got emails from heavy metal drummers sending him messages saying, 'I would lose my arm to get these kind of abilities.' Then he said, okay you don't have to lose the arm. We'll develop a third arm."

The third arm is able to work better than some drummers' real arms. It moves faster and gives the beat at a constant speed with no human error. Weinberg and Barnes worked together to create an arm that would give normal drummers a third robotic arm.

"This was very challenging. It is not something I think will actually be available even in two or five or ten years. You need to make it very convenient. You need to make it work, not too heavy. There were all kind of challenges there, but as you might have seen in the video, when he put it on the drummer, it actually worked. We get signals from the music, so we know how the music sounds. Artificial intelligence can understand and improvise, but the user can also control what sound it goes to. We also have some signals from the brain that help control the movement of the arm. We know where the body is, so we know where it puts the hand. So if the drummer moves his arm to a particular set of drums, the hand will move to another one and vice versa, so there are multiple inputs. Some of them are more autonomous, and some of them are more under control of the arm."

In *Star Wars*, Luke's hand looks like a real hand. Barnes's hand still looks very prosthetic and very metal, but the progress Weinberg has made will impact the lives of amputees greatly. There is still a long way to go before we get to the *Star Wars* level of prosthetics, but Weinberg and Barnes have made a great team that is taking prosthetics in the right direction.

Obi-Wan Kenobi: *I have taught you everything I know and you have become a far greater Jedi than I could ever hope to be, but be patient Anakin. It will not be long before the council makes you a Jedi master.*

—*STAR WARS EPISODE III: REVENGE OF THE SITH* (2005)[21]

---

21 "Star Wars: Episode III – Revenge of the Sith – Movie Quotes – Rotten Tomatoes," *Rottentomatoes.com*, accessed January 7, 2019, https://www.rottentomatoes.com/m/star_wars_episode_iii_revenge_of_the_sith/quotes/.

CHAPTER 5

MOVING FROM "I'LL SLEEP WHEN I'M DEAD" TO "I'LL SLEEP WELL TONIGHT"

———

Mantis: If I touch someone who is sad, I can ease them into contentment for a short while. I can make a stubborn person compliant. But I mostly use it to help my master sleep. He lies awake at night, thinking about his project.

Drax: Do one of those on me!

Mantis: [She touches Drax's head.] Sleep.

[Drax immediately sleeps and snores loudly.]

—*GUARDIANS OF THE GALAXY VOL. 2* (2017)[22]

Like Ego, most people lie awake at night thinking about what is going on at work. Mantis, a character in *Guardians of the Galaxy* who can make people sleep for short periods of time, helps Ego fall asleep.

Similar to Mantis, Daniel Gartenberg wants people to fall into a peaceful sleep.

Daniel Gartenberg is the creator of Sonic Sleep Coach, which is an Artificial Intelligence (AI) assistant for making sleep deeper and more regenerative. The app has smart alarm clock features that wakes you up gradually in order to prevent brain fog, or what sleep researchers call, "sleep inertia." The Internet of Things (IoT) software platform empowers people to take control of their sleep, providing customized information about how to address problems sleeping. I found Gartenberg when I came across his TED Talk entitled, "The Brain Benefits of Deep Sleep—and how to get more of it." You can tell he is passionate about his work, and he is using it to better people's lives.

---

22 "Guardians Of The Galaxy Vol. 2 – Wikiquote," *En.Wikiquote. org*, accessed January 13, 2019, https://en.wikiquote.org/wiki/ Guardians_of_the_Galaxy_Vol._2.

"I'm from Jersey. I was an undergrad at the University of Wisconsin. I was a part of a brainstorming company, and one of the ideas was an alarm clock that tracks your sleep. We ended up winning this School's Prize for Innovation and Creativity at the University of Wisconsin engineering competition. Inspired by the idea that sleep could be more regenerative, I took a neuroscience for sleep course and got into cognitive science. I ended up getting a PhD in cognitive science at George Mason University where I specialized in applied cognition and how to model fatigue using artificial intelligence.

"I started making all these algorithms to try and detect and improve sleep. I guess what inspired me was the idea that sleep is literally the human behavior we do the most. So if you could just improve sleep a little bit, it would have massive global health implications. And the science proves this. Just a small 3-5 percent improvement in deep sleep can improve next day memory performance by as much as 40 percent.

* * *

"I was big the Quantified Self movement earlier in my career. We all thought at the time that this new wave of wearable technology would change healthcare and the way we work. So I built an app around the idea that people could enter their health information and get feedback on customized ways that they could improve themselves. But we found that the time cost of people entering their information manually was too high. This high bar for entry often makes it not viable for 90 percent of the population. Maybe a few techie people would want to keep track of themselves, but it's limited. That experience made me want to noninvasively track and optimize people, and then augment that to a few questions, nothing that takes more than a minute."

I understand why people don't want to input all this information about themselves into their phone. People barely have time to sleep, let alone input their sleep patterns into a phone. As Noelle LaCharite mentions in my interview with her, no one wants to type on a screen or push buttons anymore.

I was also very into the quantified self movement during high school. I did a big science project on tracking stress through numbers and each day a few of my friends and I would input the amount of stress we had on a scale of 1 to 10. I remember some days everyone forgot and that is exactly Gartenberg's point. It's almost impossible to remember every day what you ate, how long you worked out, or how stressed you are. Noninvasively tracking people will fix that issue.

* * *

Like LaCharite, Gartenberg thinks smart home speakers, such as the Amazon Alexa, and Google Home, are a large part of our future.

"A lot of people already use some form of sleep technology, but we're thinking in the future speaker systems and non-invasive wearables that seamlessly track you will likely be in most bed rooms. It is already happening, and I think it will continue in the next five years with Amazon, Google, or Apple products. It is sort of a double-edged sword where a lot of times the technology in the room is one of the major problems resulting in people sleeping worse than we've ever slept before. My perspective on technology is that it's just a tool, and it can be used in lots of different ways—good or bad.

"I'm not a big fan of having screens in the room, but there is still a way of tracking people accurately and with this information manipulating their sound, light, and temperature environment to improve sleep. The home speaker would be an output device. But it could be a tracker too that measures sounds and breathing in the room. We already have algorithms to help us measure breathing from smart phones in the room. It is not as accurate as wearable algorithms on the Apple Watch, but it gives us some information about the sleep environment. So we are building out a system that works with all the best devices, like the Apple Watch, Oura Ring, Fitbit, Google Wear, and smart home speakers to accurately measure sleep and improve it by providing interventions that sculpt sleep into being deeper and richer in Rapid Eye Movement (REM)."

* * *

As a student, I constantly hear people bragging about how little sleep they get or how they just don't have the time. I try to get six to seven hours, which is still below the recommended level for a teenager. According to Nationwide Children's Hospital, teenagers need 9.25 hours of sleep a night and most get around seven.[23] I don't really have a choice though; I mean from school to volleyball practice to writing this book I don't have a lot of time on my hands my senior year of high school to sleep.

Gartenberg blames an overworked society for people thinking it's okay to sleep so little.

"It's a sad state of affairs when people brag about sleep depriving themselves. To me, when I think about it, it's like someone bragging about smoking cigarettes or eating a really unhealthy meal. People say things like, 'I'll sleep when I'm dead.' I think this is an example of our overworked society. It is so competitive now, especially for the middle class, that it has ended up becoming more socially acceptable for people to sleep deprive themselves. We are sleeping an hour less than we did in the 1950s when the middle class was way stronger than it is now."

---

23 "Sleep in Adolescents," *Nationwide Children's*, accessed January 10, 2019, https://www.nationwidechildrens.org/specialties/sleep-disorder-center/sleep-in-adolescents

After addressing the problem with our society and people sleeping so little, I wanted to know how Gartenberg felt knowing he had helped a lot of people sleep more and live a better life. His app has reached so many people.

"I mean, it's everything from people with tinnitus to people who have problems falling asleep and staying asleep. The software we are creating can even be useful for people who sleep with a disruptive or snoring partner in the room. One of the simplest ways to improve sleep quality is by blocking out disruptive sounds in your environment, like snoring or city sounds. I've even seen air conditioning turning on that can wake up your brain without you being consciously aware of it. So we block out noise pollution with pink and white noise that adapts to that sounds that are detected in our environment and then enhance our sleep with deep sleep stimulation sounds.

"The other thing we do the right way is wake you up. We have nailed the right way to use an alarm, though waking up naturally is always best. I also do free consultations that we plan to eventually supplant with an artificial intelligence system. Basically, almost every person I have talked to is really responsive and really loves the simple solutions we provide that can improve sleep. Sometimes it includes things in the current app, but a lot of it is based on using the best technology that's out there to address their specific problem. Some people have anxiety issues, and menopausal women have hot flashes. We're doing things to adjust for that. A lot of times the interventions involve getting people to learn how to let go because again, people have that American mentality of working all the time.

"I try to get people to think about sleep differently than they think about things like exercising because success at sleep functions very differently than what is involved in pushing yourself to your limits. The more you try to exercise, the better the results. The problem with sleep is it's not something you push. It's more about letting go. It's like the more you try to fall asleep the harder it is to fall asleep. It's teaching people how to let go."

* * *

Even with the success of his app, Gartenberg still speaks to people weekly to give them personal advice. This really shows that he cares significantly about people's sleep and his job. Sleep is something you can't practice, and I understand exactly what he is saying about having people let go. No one can sleep when they're stressed. I love that Gartenberg's work will make people more productive with more sleep and will teach them how to let go of their stress that may be making them overwork.

Gartenberg has been conducting sleep research for a long time. He loves it, and he is fascinated with the way the brain works while people are sleeping.

"I mean, when you actually look at someone's brain waves when they're sleeping, it's sort of a magical experience. You see their brain oscillating in these different stages. The brainwaves totally change. The brainwaves change with a lot of variability. And then seeing the spindles of the brainwaves and seeing the rapid eye movement; it's just kind of crazy.

"One of the things that always surprises me is that we would play hundreds of sounds at night to people, louder than how I am talking to you right now, and they would have absolutely no conscious awareness that we were playing sounds. One thing I realized was how sounds can really be disruptive when people sleep without them realizing it because we would play the sounds, and the brain basically woke up. But they would have no conscious awareness that we played the sound the next day.

"A lot of this research actually comes from hospital environments, which are noisy, and the insurance companies are worried that there will be a false alarm and an alarm goes off and someone dies. Something that surprised me a little bit is how a lot of these big companies that make most of these major devices have sleep trackers that are still pretty inaccurate after all this time. They are good at sleep wakes, but they are not that accurate at measuring sleep stages."

Gartenberg's work is applicable to everyone, which is the thing I love most about it. Everyone sleeps, and most people don't get enough sleep. One person at a time, Gartenberg is trying to change that. He is using AI to track sleep and better the life of the person using the app. This is the future, which ties back to the idea Tom Gruber brought up in his TED Talk called humanistic AI: using AI to help people. This is exactly what Gartenberg's app is doing—using technology to better human lives.

Yondu: *That's Vorker's eye. He takes it out when he sleeps. Go, look again.*

Rocket: *But leave the eye here!*

Yondu: *What?*

Rocket: *[trying not to laugh] He's gonna wake up tomorrow... AND HE'S NOT GONNA KNOW WHERE HIS EYE IS!*

—*GUARDIANS OF THE GALAXY VOL. 2 (2017)*[24]

24 "Yondu Udonta/Quote," *Marvel Cinematic Universe Wiki*, accessed January 13, 2019, http://marvelcinematicuniverse.wikia.com/wiki/Yondu_Udonta/Quote.

PART 3

SUPER AI

CHAPTER 6

HUMANISTIC AI

———

C-3PO: Excuse me sir, but that R2-D2 is in prime condition, a real bargain.

—STAR WARS EPISODE IV: A NEW HOPE (1977)[25]

C-3PO is a gold human-esque droid that I believe is a vital character in the *Star Wars* trilogy. To many fans, he is more than just a robot. 3PO is a translator, communicator, and speaks multiple languages to understand all types of creatures from wookiees to humans.

Much like 3PO, we do have a voice inside our phones that responds to commands, questions, and can Google search for

———
25 "View Quote ... Star Wars Episode IV: A New Hope ... Movie Quotes Database."

us. Siri is no 3PO, but Siri did change the way many people's worlds functioned. Siri can check the weather, Google a fact, or tell you what time is, but Siri can also send messages and call friends for people who could not communicate without the voice technology.

Tom Gruber is the founder and creator of Siri Inc.[26] Gruber says that he and his team created Siri as humanistic AI, meaning it was created to assist the needs of humans.[27] Siri works with humans and collaborates with us to make tasks simpler. Apple Inc. bought Siri in April 2010 and made it the voice technology in all Apple devices.

For most people, Siri helps us with the little things. We can ask for directions instead of typing in the address or have Siri call our favorite restaurant instead of dialing the number ourselves. However, for other people, Siri has impacted lives greatly in a variety of ways.

* * *

---

26 Tom Gruber, "How AI Can Enhance Our Memory, Work And Social Lives," posted 2017, accessed January 10, 2019, mp4 format, https://www.ted.com/talks/tom_gruber_how_ai_can_enhance_our_memory_work_and_social_lives

27 Gruber, "How AI Can Enhance Our Memory, Work And Social Lives."

During a TED Talk Gruber gave in April 2017, he tells a story about a man he knows named Daniel; Siri gave Daniel a vibrant social life.

"For my friend Daniel, the impact of the AI in these systems is a life changer. You see, Daniel is a really social guy, and he's blind and quadriplegic, which makes it hard to use those devices that we all take for granted. The last time I was at his house, his brother said, 'Hang on a second, Daniel's not ready. He's on the phone with a woman he met online.' I'm like, 'That's cool. How'd he do it?' Well, Daniel uses Siri to manage his own social life—his email, text and phone—without depending on his caregivers. This is kind of interesting, right? The irony here is great. Here's the man whose relationship with AI helps him have relationships with genuine human beings. And this is humanistic AI."[28]

Prior to humanistic AI like Siri, Daniel had to depend on others to communicate, whether it be through text or Facebook. With the creation of Siri, Daniel works with AI to give himself more social opportunities. He no longer has to rely on a caretaker or family member to help him talk to other people. Siri takes care of that for him. Humanistic AI was made to help better people's lives no matter how big or small. For me, that means having Siri set an alarm, so I know when

28 Ibid.

to move my laundry from the washer to the dryer. For Daniel, that means being able to communicate on his own.

* * *

"Now let's see where this idea of humanistic AI might lead us if we follow it into the speculative beyond. What's a kind of augmentation that we would all like to have? Well, how about cognitive enhancement? Instead of asking, 'How smart can we make our machines?' let's ask, 'How smart can our machines make us?'"[29]

This quote really struck me. In movies and in publications, I always read about how technology was going to "take over" the world. This was the first time I had ever heard someone talk about technology assisting the world. It was eye-opening to me, and I realized that it was already happening. My Amazon Alexa wakes me every day with a nice alarm. My iPhone allows me to video chat with my dad while he is away on work. Gruber continues to give examples of how technology could help us that I never thought imaginable.

Another way Gruber communicates the importance of humanistic AI is through memory.

---

29 Ibid.

"Human memory is famously flawed. We're great at telling stories, but not getting the details right. And our memories decay over time. I mean, like, where did the sixties go, and can I go there, too?"[30]

My brother always tells this story about a time he "almost got arrested." In Massachusetts, you have a passenger restriction for six months after you receive your license, which means you can only drive immediate family members. My brother was pulled over with a friend in the car and a policeman almost brought him into the station; however, my brother was lucky enough to receive a warning instead. My brother loves to tell this story at family functions, birthday parties, and pretty much any gathering. It does make people laugh, but my dad and I always notice that he is constantly changing the details to make the story a little more insane each time. This happened about three years ago now. My brother now tells the story with his friend running out of his house with knives, handcuffs, and his friend in the car crying—none of which were in the original story.

Using humanistic AI to help with memory would give me the opportunity to fact check my brother every time he tells that story. With the assistance of humanistic AI, all of one's

---

30 Ibid.

memories could be stored somewhere perfectly without being touched until you wanted to relive them.

"What if you could remember every person you ever met, how to pronounce their name, their family details, their favorite sports, the last conversation you had with them? If you had this memory all your life, you could have the AI look at all the interactions you had with people over time and help you reflect on the long arc of your relationships. What if you could have the AI read everything you've ever read and listen to every song you've ever heard? From the tiniest clue, it could help you retrieve anything you've ever seen or heard before. Imagine what that would do for the ability to make new connections and form new ideas.[31]

"What if we could remember the consequences of every food we eat, every pill we take, every all-nighter we pull? We could do our own science on our own data about what makes us feel good and stay healthy. And imagine how this could revolutionize the way we manage allergies and chronic disease."[32]

Humanistic AI has already opened doors for many people like Daniel who use AI to interact with real people. It also could revolutionize memory and potentially give someone

<hr>

31 Ibid.
32 Ibid.

a perfect memory system from the moment they are born until their last day on Earth.

A perfect memory system would allow you to remember what it felt like to eat food puree at three months old to that crazy night with you friends in your twenties and your favorite family vacation you took when you were six years old to Disney World. All your memories would be crystal clear, allowing for you to reflect on your life with true experiences.

* * *

Although we still don't have technology like C-3PO assisting battle strategy against the Empire, Gruber, along with other humanistic AI engineers, are working hard toward a reality where AI could be even more impactful and helpful than the golden robot from a galaxy far, far away.

[Chewbacca has put together C-3PO wrong after he was torn apart on the Cloud City in Empire.]

C-3PO: *Wait. Oh my! What have you done? I'm backward, you filthy furball.*
—STAR WARS EPISODE V: THE EMPIRE STRIKES BACK (1980)[33]

---

33 "View Quote ... Star Wars Episode V: The Empire Strikes Back... Movie Quotes Database."

CHAPTER 7

HOW TO PREVENT AI FROM OVERPOWERING YOU

———

Morpheus: What is real? How do you define 'real'? If you're talking about what you can feel, what you can smell, what you can taste and see, then 'real' is simply electrical signals interpreted by your brain.

—*THE MATRIX* (1999)[34]

---

34 "The Matrix: Quotes About Version Of Reality Page 2," *Schmoop.com*, accessed January 13, 2019, https://www.shmoop.com/the-matrix/versions-of-reality-quotes-2.html.

In *The Matrix*, sentient machines have taken over the human race, and perceived reality is a simulation far from that. There is a large fear technology will overpower the human race. This fear is something instilled in people today.

Max Tegmark, founder of the Future Life Institute, discusses in his TED Talk called "How to get empowered, not overpowered, by AI," how to use rising artificial intelligence to help humans.[35]

The Future Life Institute encourages beneficial use of technology. Tegmark is a professor at MIT and is a cosmologist and physicist. He is also a supporter of the effective altruism movement, which uses reasoning to determine the best way of benefitting others.

* * *

Tegmark discusses the progressions of human intelligence and technology.

"I think of the earliest life as 'Life 1.0' because it was really dumb, like bacteria, unable to learn anything during its

---

35 Max Tegmark, "How to get empowered, not overpowered, by AI," posted 2018, accessed January 10, 2019, mp4 format, https://www.ted.com/talks/max_tegmark_how_to_get_empowered_not_over-powered_by_ai?language=en

lifetime. I think of us humans as 'Life 2.0' because we can learn, which we in nerdy, geek speak, might think of as installing new software into our brains, like languages and job skills. 'Life 3.0,' which can design not only its software but also its hardware, of course doesn't exist yet. But perhaps our technology has already made us 'Life 2.1,' with our artificial knees, pacemakers and cochlear implants."[36]

AI and technology has come very far since its start in the 1950s. Not only do we have robots, but now they can walk and talk. We have self-driving cars, self-flying rockets, and drone delivery. I really liked Tegmark's analogy because it focuses on how technology helps us. Life 2.1 shows that technology is advancing humanity.

"AI really begs the question: How far will it go? I like to think about this question in terms of this abstract landscape of tasks, where the elevation represents how hard it is for AI to do each task at the human level, and the sea level represents what AI can do today. The sea level is rising as AI improves, so there's a kind of global warming going on here in the task landscape. And the obvious takeaway is to avoid careers at the waterfront, which will soon be automated and disrupted.[37]

---

36 Tegmark, "How to get empowered, not overpowered, by AI,"
37 Ibid.

"But there's a much bigger question as well. How high will the water end up rising? Will it eventually rise to flood everything, matching human intelligence at all tasks. This is the definition of artificial general intelligence—AGI, which has been the holy grail of AI research since its inception. By this definition, people who say, 'Ah, there will always be jobs that humans can do better than machines,' are simply saying that we'll never get AGI. Sure, we might still choose to have some human jobs or to give humans income and purpose with our jobs, but AGI will in any case transform life as we know it with humans no longer being the most intelligent.[38]

"Now, if the water level does reach AGI, further AI progress will be driven mainly not by humans but by AI, which means there's a possibility that further AI progress could be way faster than the typical human research and development timescale of years, raising the controversial possibility of an intelligence explosion where recursively self-improving AI rapidly leaves human intelligence far behind, creating what's known as superintelligence."[39]

I have always believed technology will never be able to accomplish some jobs, mainly those involving creativity,

---

38   Ibid.
39   Ibid.

such as interior design, painting, even entrepreneurship. All of these have to do with taste and artistic choice. Even if AI is creating products for humans, I believe humans will always understand humans better than technology.

In my opinion and most others, neither AGI nor superintelligence is in our near future. Renowned AI researcher Rodney Brooks, founder of iRobot and professor at MIT, thinks this won't happen for hundreds of years. However, a few think this could happen much sooner. Some believe we could reach AGI within decades.

* * *

Tegmark goes on to discuss what will happen if technology can do everything more efficiently than us.

"What do we want the role of humans to be if machines can do everything better and cheaper than us? The way I see it, we face a choice. One option is to be complacent. We can say, 'Oh, let's just build machines that can do everything we can do and not worry about the consequences. Come on, if we build technology that makes all humans obsolete, what could possibly go wrong?' But I think that would be embarrassingly lame.[40]

---

40 Ibid.

"I think we should be more ambitious—in the spirit of TED. Let's envision a truly inspiring high-tech future and try to steer toward it. This brings us to the second part of our rocket metaphor—the steering. We're making AI more powerful, but how can we steer toward a future where AI helps humanity flourish rather than flounder?

"To help with this, I cofounded the Future of Life Institute. It's a small nonprofit promoting beneficial technology use, and our goal is simply for the future of life to exist and to be as inspiring as possible. You know, I love technology. Technology is why today is better than the Stone Age. And I'm optimistic we can create a really inspiring high-tech future… if—and this is a big if—if we win the wisdom race, the race between the growing power of our technology and the growing wisdom with which we manage it. But this is going to require a change of strategy because our old strategy has been learning from mistakes. We invented fire, screwed up a bunch of times, and invented the fire extinguisher."[41]

I think Tegmark really hammers home a point that's vital to the future of technology; it should be used for beneficial use. Movies, TV Shows, and other media have led most to believe technology will ruin us. The Future Life Institute's goal is something that I want to be a part of and believe in.

---

41   Ibid.

As Tegmark mentions, we need to find a new way to go about inventions of technology so we can win the wisdom race.

"I have not failed. I've just found ten thousand ways that won't work," as Thomas Edison once said.

Testing all possibilities until finding the right one worked fine for Edison; however, with powerful technology like AI, learning from mistakes seems like a lousy strategy. Tegmark wants to plan ahead for all possible outcomes of advancing AI.

"It's much better to be proactive rather than reactive; plan ahead and get things right the first time because that might be the only time we'll get. But it is funny because sometimes people tell me, 'Max, shhh, don't talk like that. That's Luddite scaremongering.' But it's not scaremongering. It's what we at MIT call safety engineering. Think about it. Before NASA launched the Apollo 11 mission, they systematically thought through everything that could go wrong when you put people on top of explosive fuel tanks and launch them somewhere where no one could help them. And there was a lot that could go wrong. Was that scaremongering? No. That was precisely the safety engineering that ensured the success of the mission, and that is precisely the strategy I think we should take with AGI. Think through what can go wrong to make sure it goes right."[42]

---

42 Ibid.

＊ ＊ ＊

In this spirit, groups of AI researchers have come together
to discuss how to keep AI beneficial. They produced a list of
twenty-three principles, which have been signed by over one
thousand AI researchers.[43] Tegmark tells us about three of
the most important ones.

"One is that we should avoid an arms race and lethal autonomous weapons. The idea here is that any science can be
used for new ways of helping people or new ways of harming
people. For example, biology and chemistry are much more
likely to be used for new medicines or new cures than for
new ways of killing people because biologists and chemists
pushed hard—and successfully—for bans on biological and
chemical weapons.[44]

"In the same spirit, most AI researchers want to stigmatize
and ban lethal autonomous weapons. Another Asilomar
AI principle is that we should mitigate AI-fueled income
inequality. I think that if we can grow the economic pie dra-
matically with AI, and we still can't figure out how to divide
this pie so that everyone is better off, then shame on us."[45]

---

43 Ibid.
44 Ibid.
45 Ibid.

When thinking about creating new technology, the idea of it hurting people like biological and chemical weapons never crossed my mind. Technology is powerful though, and it is used by so many people around the world. With anything utilized by so many people, there needs to be safety protocols.

The second principle discusses the importance of AI safety. Technology has helped us, but it has also harmed the world in many ways. Tesla has autonomous driving, but this has led to many fatalities. Tegmark believes it is important to invest time and money into researching AI safety.

"We should invest much more in AI safety research because as we put AI in charge of even more decisions and infrastructure, we need to figure out how to transform today's buggy and hackable computers into robust AI systems that we can really trust because otherwise, all this awesome new technology can malfunction and harm us, or get hacked and be turned against us.[46]

"And this AI safety work has to include work on AI value alignment because the real threat from AGI isn't malice, like in silly Hollywood movies, but competence—AGI accomplishing goals that just aren't aligned with ours. For example, when we humans drove the West African black rhino extinct, we didn't do it because we were a bunch of evil rhinoceros haters. Did we? We

46 Ibid.

did it because we were smarter than them and our goals weren't aligned with theirs. But AGI is by definition smarter than us, so to make sure we don't put ourselves in the position of those rhinos if we create AGI, we need to figure out how to make machines understand our goals, adopt our goals and retain our goals."[47]

* * *

AI researchers have thoroughly discussed the outcome of AI. Tegmark explains to what the "options" are and why the outcomes are so highly debated.

"One option that some of my AI colleagues like is to build super-intelligence and keep it under human control, like an enslaved god, disconnected from the internet and used to create unimaginable technology and wealth for whoever controls it.[48]

"But Lord Acton warned us that power corrupts, and absolute power corrupts absolutely, so you might worry that maybe we humans just aren't smart enough, or wise enough rather, to handle this much power. Also, aside from any moral qualms you might have about enslaving superior minds, you might worry that maybe the superintelligence could outsmart us, break out and take over."[49]

---

47 Ibid.
48 Ibid.
49 Ibid.

I thought this idea was interesting, but it also sounds like AI is a person that is smarter than us. I personally don't think this idea would be like "an enslaved god," but I also do not believe it is the ideal outcome.

"I also have colleagues who are fine with AI taking over and even causing human extinction, as long as we feel the AIs are our worthy descendants, like our children. But how would we know the AIs have adopted our best values and aren't just unconscious zombies tricking us into anthropomorphizing them? Also, shouldn't those people who don't want human extinction have a say in the matter, too?[50]

"Now, if you didn't like either of those two high-tech options, it's important to remember that low-tech is suicide from a cosmic perspective because if we don't go far beyond today's technology, the question isn't whether humanity is going to go extinct, merely whether we're going to get taken out by the next killer asteroid, supervolcano or some other problem that better technology could have solved."[51]

The point Tegmark brings up here is not one often talked about. People always joke that everyone is always on their phones these days and that people don't interact in person

---

50 Ibid.
51 Ibid.

anymore, or they discuss the terrible effects of social media on youth. Something most people don't talk about is how much progress technology has made for us in the last century. From new medical tools to research on global warming, technology is going to save us. Without it, we would still be cave men awaiting that next killer asteroid Tegmark talks about.

* * *

Tegmark's last possible outcome is called friendly AI, which means the AI and humans have aligned morals and goals.

"So, how about having our cake and eating it ... with AGI that's not enslaved but treats us well because its values are aligned with ours? This is the gist of what Eliezer Yudkowsky has called 'friendly AI,' and if we can do this, it could be awesome. It could not only eliminate negative experiences like disease, poverty, crime and other suffering, but it could also give us the freedom to choose from a fantastic new diversity of positive experiences—basically making us the masters of our own destiny." [52]

This is similar to Gruber's idea of Humanistic AI. It allows for AI to help us and believe in the same values as humans to

---

52 Ibid.

contribute positively to human life. Friendly AI could open up a variety of doors to new experiences.

Ultimately, Tegmark explains that AGI could come soon; we have to be prepared, so we don't end up like the characters in *The Matrix*. Not planning for AI could lead to dictatorship, extinction, inequality, and more horrible outcomes.

As seen in *Avengers: Age of Ultron*, it could also lead to an intelligence that wants to wipe out the human population. With all the bad possible outcomes, Tegmark believes if we plan right, we could have every good possible outcome we wanted with the help of friendly AI.

"Do you folks want the future that's politically right or left? Do you want the pious society with strict moral rules, or do you want a hedonistic free-for-all, more like Burning Man 24/7? Do you want beautiful beaches, forests and lakes, or would you prefer to rearrange some of those atoms with the computers, enabling virtual experiences? With friendly AI, we could simply build all of these societies and give people the freedom to choose which one they wanted to live in because we would no longer be limited by our intelligence, merely by the laws of physics. So the resources and space for this would be astronomical—literally.[53]

---

53   Ibid.

"We're all here to celebrate the age of amazement, and I feel that its essence should lie in becoming not overpowered but empowered by our technology."[54]

Tony Stark: JARVIS, are you up?

JARVIS: For you, sir, always.

Tony Stark: I'd like to open a new project file, index as: Mark 2.

JARVIS: Shall I store this on the Stark Industries' *central database?*

Tony Stark: I don't know who to trust right now. 'Til further notice, why don't we just keep everything on my private server.

JARVIS: Working on a secret project, are we, sir?

Tony Stark: I don't want this winding up in the wrong hands. Maybe in mine it could actually do some good.

—IRON MAN (2008)[55]

54 Ibid.

55 "J.A.R.V.I.S./Quote," *Marvel Cinematic Universe Wiki*, accessed January 13, 2019, http://marvelcinematicuniverse.wikia.com/wiki/J.A.R.V.I.S./Quote.

CHAPTER 8

AI CAN FIGHT ONLINE HARASSMENT AND EXTREMISM

───

Ultron: Shhhh. I'm here to help. [Ultron starts absorbing Jar-vis's consciousness.]

JARVIS: Stop! Please...may I...I...! I cannot...cannot... [Ultron then begins to prepare himself a body from body parts of the Iron Legion.] [Meanwhile, the Avengers mingle at the party.]

—*AVENGERS: AGE OF ULTRON* (2015)[56]

───────────

56 "J.A.R.V.I.S./Quote."

From *Transformers* to *Star Wars*, viewers can see how technology helps a hero or creates a hero. But, we also see movies like *Avengers: Age of Ultron*, or even *Despicable Me*, where technology is a significant aid to the enemy.

"My relationship with the internet reminds me of the setup to a cliché horror movie. You know, the blissfully happy family moves in to their perfect new home, excited about their perfect future. It's sunny outside and the birds are chirping … And then it gets dark. And there are noises from the attic. And we realize that that perfect new house isn't so perfect."[57]

This is the opening lines from Yasmin Green's TED Talk in 2018. Green started working for Google in 2006, and since then, has become relatively well-known for her work. *Vogue* has even written about her. Green grew up in Iran, and her family moved to London in 1984.[58]

* * *

Green works for Jigsaw, which as described by Robert Sullivan a writer for *Vogue*, is "the tech giant's in-house think tank—and the trouble at hand has to do with trolls, those who foment anger

57 Yasmin Green, "How technology can fight online harassment and extremism," posted 2018, accessed January 10, 2019, mp4 format, https://www.ted.com/talks/yasmin_green_how_technology_can_fight_extremism_and_online_harassment

58 Green, "How technology can fight online harassment and extremism."

and disrupt discourse online. People tend to think of trolls in terms of random harassment teenagers with too much time on their hands, but these in particular are globally scattered and state-sponsored, targeting journalists and activists, independent press, or outspoken citizens."[59] Green describes her job as getting better information into the hands of vulnerable people.

When Green started working in tech, she was in awe of the internet.[60] Facebook was only two years old, and Twitter didn't even exist yet. Green was in love with the idea of technology. However, since the beginning, she saw negative aspects of the internet.

"We [Green and her colleagues] were doing the inspiring work of building search engines and video-sharing sites and social networks. Criminals, dictators and terrorists were figuring out how to use those same platforms against us. And we didn't have the foresight to stop them.

"Over the last few years, geopolitical forces have come online to wreak havoc. And in response, Google supported a few colleagues and me to set up a new group called Jigsaw, with a mandate to make people safer from threats like violent extremism, censorship,

---

59 Sullivan Robert, "Meet the Head International Troll Slayer at Google," *Vogue*, accessed January 10, 2019, https://www.vogue.com/article/google-jigsaw-yasmin-green-internet-trolls-web-security

60 Green, "How technology can fight online harassment and extremism."

persecution—threats that feel very personal to me because I was born in Iran, and I left in the aftermath of a violent revolution. But I've come to realize that even if we had all of the resources of all of the technology companies in the world, we'd still fail if we overlooked one critical ingredient—the human experiences of the victims and perpetrators of those threats."[61]

* * *

During this talk, Green takes us through two major issues. The first is terrorism. She tells a story about a British girl going to Syria to join ISIS. She was only thirteen years old, and her father supported her.

"I said, 'Why?' And she said, 'I was looking at pictures of what life is like in Syria, and I thought I was going to go and live in the Islamic Disney World.' That's what she saw in ISIS. She thought she'd meet and marry a jihadi Brad Pitt and go shopping in the mall all day and live happily ever after."[62]

Green goes on to explain how insanely smart ISIS is with their marketing.

---

61   Ibid.
62   Ibid.

"ISIS understands what drives people, and they carefully craft a message for each audience. Just look at how many languages they translate their marketing material into. They make pamphlets, radio shows and videos in not just English and Arabic, but German, Russian, French, Turkish, Kurdish, Hebrew, Mandarin Chinese. I've even seen an ISIS-produced video in sign language.[63]

"Just think about that for a second: ISIS took the time and made the effort to ensure their message is reaching the deaf and hard of hearing. Tech-savviness is not the reason ISIS wins hearts and minds. Their insight into the prejudices, the vulnerabilities, the desires of the people they're trying to reach wins them over. That's why it's not enough for the online platforms to focus on removing recruiting material. If we want to have a shot at building meaningful technology that's going to counter radicalization, we have to start with the human journey at its core."[64]

* * *

Green's team went to talk to young members of ISIS who had bought into ISIS's marketing and their motivations but eventually realized they had made a mistake.

---

63 Ibid.
64 Ibid.

"I'm sitting there in this makeshift prison in the north of Iraq with this twenty-three-year-old who had actually trained as a suicide bomber before defecting. And he says, 'I arrived in Syria full of hope, and immediately, I had two of my prized possessions confiscated: my passport and my mobile phone.' The symbols of his physical and digital liberty were taken away from him on arrival. This is the way he described that moment of loss to me. He said, 'You know in 'Tom and Jerry,' when Jerry wants to escape, and then Tom locks the door and swallows the key and you see it bulging out of his throat as it travels down?' And of course, I really could see the image that he was describing, and I really did connect with the feeling he was trying to convey, which was one of doom, when you know there's no way out."[65]

The most terrifying part of this was that when Green asked if there was anything that could have changed his mind the day he left, he said no.

"At that point, I was so brainwashed, I wasn't taking in any contradictory information. I couldn't have been swayed."[66]

---

65 Ibid.
66 Ibid.

However, he went on to say that if he knew everything he knew now, six months before his departure, it would have made him stay at home.

"Radicalization isn't this yes-or-no choice. It's a process, during which people have questions—about ideology, religion, living conditions. And they're coming online for answers, which is an opportunity to reach them. There are videos online from people who have answers—defectors, for example, telling the story of their journey into and out of violence; stories like the one from that man I met in the Iraqi prison. There are locals who've uploaded cell phone footage of what life is really like in the caliphate under ISIS's rule. There are clerics who are sharing peaceful interpretations of Islam. But you know what? These people don't generally have the marketing prowess of ISIS. They risk their lives to speak up and confront terrorist propaganda, and then they tragically don't reach the people who most need to hear from them. And we wanted to see if technology could change that."[67]

I personally have never come across ISIS marketing videos, but I have also never seen the people standing up against ISIS. Maybe none of these videos reach me for a reason, but the people confronting terrorist propaganda should be seen by everyone.

---

67   Ibid.

* * *

In 2016, Jigsaw partnered with Moonshot CVE to create a new approach to countering radicalization called "Redirect Method."[68]

"It uses the power of online advertising to bridge the gap between those susceptible to ISIS's messaging and those credible voices that are debunking that messaging. It works like this: someone looking for extremist material—say they search for 'How do I join ISIS?'—will see an ad appear that invites them to watch a YouTube video of a cleric, of a defector—someone who has an authentic answer. And that targeting is based not on a profile of who they are, but of determining something that's directly relevant to their query or question."[69]

During their eight-week pilot in English and Arabic, they reached over three hundred thousand people who had shown interest in a Jihadi group. The Redirect Method shows people videos of those standing up against ISIS. Green says the Redirect Method is being used globally to protect people from a variety of extremist groups and gives many people the opportunity to choose differently.

---

68 Ibid.
69 Ibid.

"It turns out that often the bad guys are good at exploiting the internet, not because they're some kind of technological geniuses, but because they understand what makes people tick.

"I want to give you a second example—online harassment. Online harassers also work to figure out what will resonate with another human being, not to recruit them like ISIS does but to cause them pain.

"Imagine this. You're a woman, you're married, you have a kid. You post something on social media, and in a reply, you're told that you'll be raped, that your son will be watching, details of when and where. In fact, your home address is put online for everyone to see. That feels like a pretty real threat. Do you think you'd go home? Do you think you'd continue doing the thing you were doing? Would you continue doing that thing that's irritating your attacker?"[70]

Green goes on to talk about how online abusers know what makes people afraid, insecure, and angry.

"When online harassment goes unchecked, free speech is stifled. And even the people hosting the conversation throw up their arms and call it quits, closing their comment sections

_____

70 Ibid.

and their forums altogether. That means we're actually losing spaces online to meet and exchange ideas. And where online spaces remain, we descend into echo chambers with people who think just like us. But that enables the spread of disinformation; that facilitates polarization. What if technology instead could enable empathy at scale?"[71]

* * *

Green and her team partnered with Google's Counter Abuse team, Wikipedia, and newspapers to see if they could build machine-learning models that could understand the emotional impact of language.

"Could we predict which comments were likely to make someone else leave the online conversation? And that's no mean feat. That's no trivial accomplishment for AI to be able to do something like that. I mean, just consider these two examples of messages that could have been sent to me last week. 'Break a leg at TED!' … and 'I'll break your legs at TED.'"[72]

People laugh to this joke, but the idea of it is fascinating. On a smaller scale, this could end teenage cyber-bullying all together. Instead of having people report cyber bullying and

---

71 Ibid.
72 Ibid.

hate posts on social media, AI could immediately recognize it and delete it before it affects someone. Eighty-seven percent of today's youth say they have witnessed cyber bullying.[73] Green's team is attempting to find a way to eliminate hate speech online. Right now, technology doesn't process words and tone of voice, but with Green's team's developments, it's starting to.

"The beauty of building AI that can tell the difference is that AI can then scale to the size of the online toxicity phenomenon, and that was our goal in building our technology called Perspective. With the help of Perspective, *The New York Times*, for example, has increased spaces online for conversation. Before our collaboration, they only had comments enabled on 10 percent of their articles. With the help of machine learning, they have that number up to 30 percent. So they've tripled it, and we're still just getting started."[74]

From the future possibility of AI taking over the world to online abuse and terrorism, a lot of fear abounds around technology, but Green and her team show us that technology will fix our problems. The reason I brought up movies with both beneficial and detrimental technologies at the

---

73   "Cyber-bullying Facts and Statistics," *Teen Safe*, accessed January 10, 2019, https://www.teensafe.com/blog/cyber-bullying-facts-and-statistics/

74   Ibid.

beginning of this piece was to show that it can be used both ways—something Green describes well in her speech.

"When people use technology to exploit and harm others, they're preying on our human fears and vulnerabilities. If we ever thought we could build an internet insulated from the dark side of humanity, we were wrong. If we want today to build technology that can overcome the challenges we face, we have to throw our entire selves into understanding the issues and into building solutions that are as human as the problems they aim to solve. Let's make that happen."[75]

Baymax: [to Hiro, who's stuck and buried under a pile of action figures] On a scale of one to ten, how would you rate your pain?

Hiro: [irritated] Zero.

Baymax: It is alright to cry.

Hiro: No! No, no, no, no, no!

Baymax: [picks up Hiro and holds him like a baby] Crying is a natural response to pain.

---

75   Ibid.

Hiro: [jumps out of Baymax's arms] I'm not crying.

Baymax: I will scan you for injuries.

Hiro: [firmly] DON'T scan me.

Baymax: Scan complete.

—*BIG HERO 6* (2014)[76]

---

76 "Big Hero 6 - Calgary Public Library," *Bibliocommons*, accessed January 13, 2019, https://calgary.bibliocommons.com/item/ quotation/987362095.

CHAPTER 9

TEACHING COMPUTERS TO UNDERSTAND PICTURES

———

Wanda: Vision, are you not letting me leave?

Vision: It is a question of safety.

Wanda: I can protect myself.

Vision: Not yours. Mr. Stark would like to avoid the possibility of another public incident. Until the Accords are on a... more secured foundation.

Wanda: And what do you want?

Vision: *For people to see you... as I do.*

—CAPTAIN AMERICA: CIVIL WAR (2016)[77]

JARVIS, Tony Stark's sidekick and artificial intelligence, can understand essentially anything, and he eventually turned into his own character when he became Vision.

Although we are far from JARVIS or Vision, some would like to believe we're close. I used to think we were, until I watched Fei Fei Li's TED Talk on teaching computers how to understand pictures and how to see objects.

Li has two kids—a son and a daughter—and she emigrated from China when she was sixteen.[78] Growing up, she worked multiple jobs on top of her school work because her parents did not have high-paying jobs due to the fact that her parents did not speak English.

* * *

---

77 "Vision/Quote," *Marvel Cinematic Universe Wiki*, accessed January 13, 2019, http://marvelcinematicuniverse.wikia.com/wiki/Vision/Quote.

78 Fei Fei Li, "How we're teaching computers to understand pictures," posted 2015, accessed January 13, 2019, mp4 format, https://www.ted.com/talks/fei_fei_li_how_we_re_teaching_computers_to_under-stand_pictures?language=en.

Right now, after arduous work, computers can only understand images to the level of a three-year-old human. Li is a professor of computer science at Stanford University, and she is the director of the Stanford Artificial Intelligence Lab and the Stanford Vision Lab.

For the past few years, her main focus has been trying to teach computers to understand objects from cats to birthday candles.

"Seeking knowledge and truth was in my blood. I wanted to understand the universe and I wanted that kind of intellectualism in my life."

Li explains how we have super advanced technology, but a lot of this tech still cannot understand pictures. Having these devices comprehend what they are seeing could help them perform their tasks more accurately.

"We have prototyped cars that can drive by themselves, but without smart vision, they cannot really tell the difference between a crumpled paper bag on the road, which can be run over, and a rock that size, which should be avoided. We have made fabulous megapixel cameras, but we have not delivered sight to the blind. Drones can fly over massive land, but they don't have enough vision technology to help us to track the changes of the rainforests. Security cameras are everywhere, but they do not alert us when a child is drowning in a swimming pool. Photos and videos are becoming an integral part of global life.[79]

"They're being generated at a pace that's far beyond what any human, or teams of humans, could hope to view, and you and I are contributing to that at this TED. Yet our most advanced software is still struggling at understanding and managing this enormous content. So in other words, collectively as a society, we're very much blind because our smartest machines are still blind."[80]

* * *

Li goes on to talk about how exactly her team trained computers to identify these images. Essentially, they took a series of images from the internet of one object, cats. They trained the computer to try and recognize a normal looking cat sitting up right with perfect pointy ears.

However, there is an issue with the learning process. Cats come in all shapes and sizes. They lie in different positions, and they perform different actions that make them look different. Every object, even a cat, has infinite variations. Teaching the computer to see all cats was the tricky part.

---

79 Li, "How we're teaching computers to understand pictures."
80 Ibid.

Eight years ago, Li started to compare her process to how a child learns how to understand pictures.[81]

"No one tells a child how to see, especially in the early years. They learn this through real-world experiences and examples. If you consider a child's eyes as a pair of biological cameras, they take one picture about every two hundred milliseconds, the average time an eye movement is made. So by age three, a child would have seen hundreds of millions of pictures of the real world. That's a lot of training examples. So instead of focusing solely on better and better algorithms, my insight was to give the algorithms the kind of training data that a child was given through experiences in both quantity and quality."[82]

I honestly thought that training a computer to understand objects would be easier than an infant, but as Li explains, infants see millions of images all the time. They have more exposure to every object. Teaching a computer will also take massive amounts of pictures for training.

* * *

In 2007, Li along with Professor Kai Li at Princeton launched ImageNet.[83] They went to the internet, where they

---

81   Ibid.
82   Ibid.
83   Ibid.

downloaded billions of images and the Amazon Mechanical Turk platform to help them label each image. At one point, ImageNet had over fifty thousand workers.[84]

By 2009, ImageNet had fifteen billion images in over twenty-two thousand classes of objects. They had over sixty-two thousand images of cats, and they opened up the data set to the world.[85]

"As it turned out, the wealth of information provided by ImageNet was a perfect match to a particular class of machine learning algorithms called convolutional neural network, pioneered by Kunihiko Fukushima, Geoff Hinton, and Yann LeCun back in the 1970s and '80s.[86]

"Just like the brain consists of billions of highly connected neurons, a basic operating unit in a neural network is a neuron-like node. It takes input from other nodes and sends output to others. Moreover, these hundreds of thousands or even millions of nodes are organized in hierarchical layers, also similar to the brain. In a typical neural network we use to train our object recognition model, it has twenty-four million nodes, one hundred forty million parameters, and fifteen billion connections. That's an enormous model. Powered by the massive data from ImageNet and the modern CPUs and

---

84 Ibid.
85 Ibid.
86 Ibid.

GPUs to train such a humongous model, the convolutional neural network blossomed in a way that no one expected. It became the winning architecture to generate exciting new results in object recognition.[87]

"Sometimes, when the computer is not so confident about what it sees, we have taught it to be smart enough to give us a safe answer instead of committing too much, just like we would do, but other times our computer algorithm is remarkable at telling us what exactly the objects are, like the make, model, year of the cars."[88]

Li and her team used their new image-recognizing computer to find out some fairly interesting trends, such as car prices are linked to crime rates in a city.

* * *

Li's team has now taught computers to recognize objects. This is similar to a toddler pointing and yelling at objects as they walk down the street. The computer is able to recognize certain objects but not fully understand the context the objects are surrounded by. Creating sentences was the

---

87 Ibid.
88 Ibid.

next step, and it would be another bridge between Big Data and machine learning algorithms.

Four years ago, Li and her team were able to teach a computer how to say sentences about an image.[89] Although she knows this is an amazing accomplishment, she also realizes that the computer still makes a lot of mistakes. In her TED Talk, she shows a video of a computer mistaking ordinary objects for other things they are not. There is still a lot of work to be done.

In the last minutes of her TED Talk, Li explains how far this program and her field still has to go.

"But there's so much more to this picture than just a person and a cake. What the computer doesn't see is that this is a special Italian cake that's only served during Easter time. The boy is wearing his favorite t-shirt given to him as a gift by his father after a trip to Sydney, and you and I can all tell how happy he is and what's exactly on his mind at that moment."[90]

Slowly, we are making machines smarter to enhance humans. We are teaching computers to see, so they can help humans see better.

---

89 Ibid.
90 Ibid.

"This is my quest: to give computers visual intelligence and to create a better future for the world."[91]

Vision: [to Wanda] It's alright. You could never hurt me. I just feel you.

—AVENGERS: INFINITY WAR (2018)[92]

91 Ibid.
92 "Vision/Quote."

CHAPTER 10

INNOVATIVE AND EXCITING JOBS WILL BE THE NEW OFFICE JOBS

—

Robert E. Lee Prewitt: A man should be what he can do.

—FROM HERE TO ETERNITY (1953)[93]

Many alternate futures in sci-fi movies are an utter dystopia. From *Ready Player One* to the *Divergent* movies, Ohio and Illinois seem like a mess by the time we reach even 2045. I watch so many sci-fi movies, sometimes I start to wonder if *The Hunger Games* or a world similar could become a reality.

93 "From Here To Eternity (1953)," *Imdb*, accessed January 13, 2019, https://www.imdb.com/title/tt0045793/characters/nm0001050.

All of these fictional societies have somewhat real aspects to them, which is why they resonate with us and these books and movies sell; however, David Lee, the Vice President of Innovation and the Strategic Enterprise Fund for UPS, has convinced me that the future of the world could actually be more productive and fun than it is today.

"You see, I work in innovation, and part of my job is to shape how large companies apply new technologies. Certainly some of these technologies are even specifically designed to replace human workers. But I believe that if we start taking steps right now to change the nature of work, we can not only create environments where people love coming to work but also generate the innovation that we need to replace the millions of jobs that will be lost to technology.[94]

"I believe the key to prevent our jobless future is to rediscover what makes us human and to create a new generation of human-centered jobs that allow us to unlock the hidden talents and passions we carry with us every day."[95]

* * *

---

94 David Lee, "Why jobs of the future won't feel like work," posted 2017, accessed January 13, 2019, mp4 format, https://www.ted.com/talks/david_lee_why_jobs_of_the_future_won_t_feel_like_work.

95 Lee, "Why jobs of the future won't feel like work."

Although Lee is excited about our technologically advanced future, he also believes that the "technology taking over the world" mindset is something we brought upon ourselves.

"But first, I think it's important to recognize that we brought this problem on ourselves. It's not just because we are the one building the robots. But even though most jobs left the factory decades ago, we still hold on to this factory mindset of standardization and de-skilling. We still define jobs around procedural tasks and then pay people for the number of hours they perform these tasks. We've created narrow job definitions like cashier, loan processor or taxi driver and then asked people to form entire careers around these singular tasks."[96]

Lee makes an interesting point here. A lot of jobs are centered around a few tasks that the person just has to do well enough on a daily basis to not be fired. It got me thinking: what if jobs were creative and innovative? Not just focused on a few small things?

* * *

Lee talks about the implications of this "singular task" mindset and how it has actually allowed people to be replaced by robots more easily.

---

96 Ibid.

"These choices have left us with actually two dangerous side effects. The first is that these narrowly defined jobs will be the first to be displaced by robots because single-task robots are just the easiest kinds to build. But second, we have accidentally made it so that millions of workers around the world have unbelievably boring working lives.[97]

"Let's take the example of a call center agent. Over the last few decades, we brag about lower operating costs because we've taken most of the need for brainpower out of the person and put it into the system. For most of their day, they click on screens and read scripts. They act more like machines than humans. And unfortunately, over the next few years, as our technology gets more advanced, they, along with people like clerks and bookkeepers, will see the vast majority of their work disappear.[98]

"To counteract this, we have to start creating new jobs that are less centered on the tasks a person does and more focused on the skills a person brings to work. For example, robots are great at repetitive and constrained work, but human beings have an amazing ability to bring together capability with creativity when faced with problems we've never seen before. When every day brings a little bit of a surprise, we have designed work for humans and not for robots. Our entrepreneurs and engineers already live in this world, but

---

97   Ibid.
98   Ibid.

so do our nurses and our plumbers and our therapists. You know, it's the nature of too many companies and organizations to just ask people to come to work and do their jobs."[99]

This made me think of the TED Talk I watched where Tom Gruber spoke about the future of AI. Humanistic AI is the idea that technology and humans can work together to benefit human life. Lee is discussing the exact same topic here. What if we used technology to enhance the workplace instead of replacing workers? Although Gruber was discussing less work and more personal life, technology can benefit almost any aspect of life if we use it in the right way.

"But if your work is better done by a robot, or your decisions better made by an AI, what are you supposed to be doing? Well, I think for the manager, we need to realistically think about the tasks that will be disappearing over the next few years and start planning for more meaningful, more valuable work that should replace it. We need to create environments where both human beings and robots thrive.[100]

"I say, let's give more work to the robots, and let's start with the work that we absolutely hate doing. For the human beings, we should follow the advice from Harry Davis at

---

99 Ibid.
100 Ibid.

the University of Chicago. He says we have to make it so that people don't leave too much of themselves in the trunk of their car. I mean, human beings are amazing on weekends. Think about the people you know and what they do on Saturdays. They're artists, carpenters, chefs and athletes. But on Monday, they're back to being Junior HR Specialist and Systems Analyst 3."[101]

* * *

In their free time, people are so interesting and innovative. At school, most kids just sit in their normal high school classes learning all the things you probably learned in high school too. Outside the classroom, some are classical pianists, synchronized swimmers, rock climbers, and bakers. According to Gallup, only fifteen percent of world's full-time workers feel engaged at work.[102]

Lee goes on to tell a story where he found that what people do on in their free time is really their main skill set and how it could be their future. Your average office worker could be building model houses with toothpicks on his/her Friday nights; you never know.

---

101 Ibid.

102 Clifton Jim, "The World's Broken Workplace," *Gallup*, accessed January 10, 2019, https://news.gallup.com/opinion/chairman/212045/world-broken-workplace.aspx?g_source=position1&g_medium=related&g_campaign=tiles

"A few years ago, I was working at a large bank that was trying to bring more innovation into its company culture. So my team and I designed a prototyping contest that invited anyone to build anything that they wanted. We were actually trying to figure out whether or not the primary limiter to innovation was a lack of ideas or a lack of talent, and it turns out it was neither one. It was an empowerment problem. And the results of the program were amazing.[103]

"We started by inviting people to re-envision what they could bring to a team. This contest was not only a chance to build anything they wanted but also be anything they wanted. And when people were no longer limited by their day-to-day job titles, they felt free to bring all kinds of different skills and talents to the problems they were trying to solve.[104]

"We saw technology people being designers, marketing people being architects, and even finance people showing off their ability to write jokes. We ran this program twice, and each time more than four hundred people brought their unexpected talents to work and solved problems they had been wanting to solve for years. Collectively, they created millions of dollars of value, building things like a better touch-tone system for call centers, easier desktop tools for branches, and

---

103 Ibid.
104 Ibid.

even a thank you card system that has become a cornerstone of the employee working experience.[105]

"Over the course of the eight weeks, people flexed muscles they never dreamed of using at work. People learned new skills, met new people, and at the end, somebody pulled me aside and said, 'I have to tell you, the last few weeks have been one of the most intense, hardest working experiences of my entire life, but not one second of it felt like work.'"[106]

When people are investing their time into creating something that is intellectually stimulating for them, not only is their product or output amazing, but they are actually enjoying themselves. This surprised me, but it also didn't at the same time. I know people who love to dance or draw, but because those fields are not as financially stable as others, their families encourage them to pursue other things. They still have those skills, but they only use them in their free time. People have skill sets they might not even believe are valuable, but Lee's point is that everyone is an innovator.

* * *

---

105 Ibid.
106 Ibid.

Using technology, Lee thinks this kind of innovation and excitement is possible as a common job when combined with technology.

"I believe the jobs of the future will come from the minds of people who today we call analysts and specialists, but only if we give them the freedom and protection they need to grow into becoming explorers and inventors. If we really want to robot-proof our jobs, we, as leaders, need to get out of the mindset of telling people what to do and instead start asking them what problems they're inspired to solve and what talents they want to bring to work.[107]

"Because when you can bring your Saturday self to work on Wednesdays, you'll look forward to Mondays more, and those feelings we have about Mondays are part of what makes us human."[108]

John Keating: Carpe Diem. Seize the day, boys. Make your lives extraordinary.

—*DEAD POET'S SOCIETY* (1989)[109]

---

107 Ibid.

108 Ibid.

109 "Robin Williams' Best Dead Poets Society Quotes: 'Carpe Diem. Seize The," *The Independent*, accessed January 13, 2019, https://www. independent.co.uk/arts-entertainment/films/news/robin-williams-best-dead-poets-society-quotes-carpe-hear-it-carpe-carpe-diem-seize-the-day-boys-9663800.html.

PART 4

VOICE TECHNOLOGY IS THE FUTURE

CHAPTER 11

WHY THE SENIOR ARCHITECT OF THE AMAZON ALEXA BELIEVES VOICE TECHNOLOGY IS THE FUTURE

———

JARVIS: Good morning. It's 7 a.m. The weather in Malibu is seventy-two degrees with scattered clouds. The surf conditions are fair with waist to shoulder highlines. High tide will be at 10:52 a.m.

—IRON MAN (2008)[110]

---

110 "J.A.R.V.I.S./Quote."

I used to think that touch screens were all the hype. They're everywhere. Tablets, computers, iPhones, Galaxys, the list goes on.

After talking to people like Noelle LaCharite and listening to lectures from Umesh Sachdev, I know that voice technology can reach people like screens can't. The immediate example is when your hands are full and you simply cannot type.

As Sachdev discusses, the lower class gains a lot from voice technology. They are able to talk to technology in any language, enabling them to perform actions, such as wire money or message a loved one, that were not possible with just screens.

Human beings are busy. Who wants to take the time to type on a screen when they can just talk to a device?

* * *

LaCharite believes that the future is voice, not screens. When the iPhone came out, it was a cultural phenomenon. I was shocked when I heard her say screens are not the future because I couldn't imagine my life without a screen, laptop, or smartphone. LaCharite was the Senior Architect on Amazon's Alexa and now works at Microsoft.

I thought she was biased toward voice over screen because she helped create Alexa. She changed my mind in a twenty-minute conversation.

"So when I started at Amazon, 'Voice First' was our motto. If you built anything, you built it without a screen, and then if you got that working and you got customers, you could build it with a screen. But since then, I guess my philosophy has evolved a little bit because I think voice should be definitely a UI [user interface] that you build for, but ultimately will be multi-motoral."

I was shocked to hear that Amazon's motto was Voice First. I think it's interesting that voice is becoming the next big thing after touch screens. The Amazon Alexa have been a huge hit and Google followed with their Google Homes. Amazon knows what they're doing.

"If I have a screen, yeah I might touch it, but I don't want to have to drop down a menu and find something right? I want to just be able to ask for what I want. But I might touch the screen or swipe the screen or use gesture control, so I don't think any of that will go away. I do think the concept of holograms, like you can imagine virtual assistants like Alexa having a holographic representation at some point where you feel like you're talking to a thing rather than kind of nothing, like a device. It could start showing up. All of these assistants could have little cartoon type graphics that show up on computers that say 'hey, it's Alexa. What do you need from me?' They're not human in any way; it'd be like a cloud or a bubble or a circle or whatever.

"I think the one reason a lot of people shy away from it is it's a hard problem to solve. Humans say things in infinite numbers of ways, and it is very difficult to design for that."

LaCharite discusses an issue many, many of my people have with Siri. My last name is Yun, but it is pronounced "Yoon." When I try to call my siblings using Siri, it almost never works. People say things in so many different ways. It would be so hard for a machine to understand all different languages, dialects, and accents.

* * *

"We're not at a level of maturity in our machine learning algorithms to be unsupervised at that level where the model can train itself on all the random things humans might say and work. So even today, it's a very manual effort. Someone is literally listening to audio and tagging it with what it means and then we use that as a rule to kind of make other decisions, but if we find that the rule is broken, we manually go back and fix it. Someone is writing that code and putting it in the model. There is no automation of that. Maybe there is somewhere, but not in these big language providers. It's just tough."

This immediately made me think about the TED Talk with Yasmin Green. She was trying to have AI understand the context of online language. Machines are not yet able to figure out and adapt to what humans are saying on their own. That's the next big step, and I think that will change everything. If a computer can learn on its own to understand a specific human, it can tailor itself to know all the little phrases and weird pronunciations a person says.

"One of the main reasons is that in science fiction there is a lot of talking to computers, which is why I was passionate about it.

"My dad taught me very well all of these classic golden age of science fiction stories and in all of them you could talk to computers. It wasn't the only thing you could do, like there was buttons and radio dials, all sorts of different UIs."

This brought me back to the first time I watched *Star Wars* and *Star Trek*. *Star Wars* was easily the biggest life inspiration I ever had. It seems like sci-fi movies had the same effect on LaCharite as they have on me.

"But, voice was one of them, and more importantly, now that I have kids, and I'm watching them interact with software, my one-year-old is saying words granted the machine can't pick it up yet, but she is definitely saying Alexa. And my three-year-old definitely knows how.

"We'll go on YouTube and hit the microphone button on YouTube and search for Garfield. Like he knows Garfield now, so he hits that button for Garfield. He won't type; he doesn't know how to write words yet. So, I would imagine since he started at an infant seeing people talk to technology.

"Our little kids today know we have phones and those phones are computers. I think voice will be very similar to little kids of today as they become engineers and customers in the workforce. They're going to expect to be able to say things to their computer."

I never really thought about it in that way. I've had an iPhone since I was around thirteen, and I still remember when the iPhone came out. I was six, and I was going into second grade. I still grew up knowing that technology was a privilege. To me, computers were always a given. They were everywhere by the time I was born, but cell phones were not a given to me. I can't imagine thinking that talking to phones and computers is normal while I was growing up, but that is what young children today think.

When my mom came home with her first iPhone, I thought it was the coolest thing ever. I can't say I lived through the 80s and 90s when the internet was becoming popular, but I forget some kids didn't know what it was like when Blackberries and flip phones were still popular.

My sister was three when the iPhone came out. She told me she doesn't remember life before the iPhone. It's like LaCharite's kids not knowing life without voice technology.

* * *

"I always tell people 'just start building now, it might be horrible, but certainly in twenty years we'll be better at it if we start today,' but a lot of people can't see that far. They can't see the value in investing in something that potentially they won't even be around to see.

"That's one of the things I really like about working at Amazon. For someone like Jeff Bezos, he was constantly saying the work you are doing today will not impact this generation. You will not be alive to see the value this is providing. It was just so hard to imagine. Everything you do is not for you, it's for your kid's kids. I just think that's such a unique way of looking at the world."

This was an interesting statement. It almost makes people who work in tech selfless in a way. They know that a lot of what they are producing will not benefit them and will only benefit those after them. I agree with LaCharite; it is such a unique way of looking at the world.

JARVIS: Test complete. Preparing to power down and begin diagnostics...

Tony Stark: *Uh, yeah, tell you what. Do a weather and ATC check. Start listening in on ground control.*

JARVIS: Sir, there are still terabytes of calculations required before an actual flight is...

Tony Stark: *JARVIS, sometimes you gotta run before you can walk.*

—*IRON MAN* (2008)[111]

---

111 "J.A.R.V.I.S./Quote."

CHAPTER 12

VOICE TECHNOLOGY WILL ENHANCE ALL LIVES

JARVIS: I'll continue to run variations on the interface, but you should probably prepare for your guests. I'll notify you if there are any developments.

Tony Stark: *Thanks, buddy.*

JARVIS: Enjoy yourself, sir.

Tony Stark: *I always do.*

—*AVENGERS: AGE OF ULTRON* (2015)[112]

"This is Radha [shows photo of woman]. She lives in a backward area of Bihar, and she wants to send Rs. 500 [the Indian currency] to her daughter for Diwali. She is not educated enough to do it through internet banking, and neither is there any banking facility close to her village.[113]

"It would have been so easy if Radha could give a command on her cellphone in her own language and the money could be transferred. Then she could receive a confirmation in her language, and it would be a happy Diwali for her family. To bring happiness into the lives of people like Radha, a technology was needed that was easy to operate and could empower the people."[114]

* * *

Umesh Sachdev, the CEO of Uniphore Software Systems, said he started his work on voice technology with this example

---

112 Ibid.

113 Umesh Sachdev, "The future of voice technology," posted 2017, accessed January 13, 2019, mp4 format, https://www.ted.com/talks/umesh_sachdev_the_future_of_voice_technology?language=en.

114 Sachdev, "The future of voice technology."

in mind. He believes technology should be used by everyone and only then it will be under humans' control.

"Just imagine if we could speak to machines the way we speak to humans, how easy would our lives be. Business would be so simple. People who fear technology would also have an easier life."[115]

AI is starting to give us the ability to talk to machines, but AI is still not even close to understanding everything. As Noelle LaCharite also discusses, AI doesn't understand every language, our tone, or what words we emphasize when we speak. Sachdev is working to change this problem.

"And now in India, our team at Uniphore has taught seventeen Indian languages and about eighty international languages to machines, making the communication between man and machine easy. We started this project ten years ago in IIT Madras and since then, many companies have used it to grow their business."[116]

* * *

115 Ibid.
116 Ibid.

Sachdev isn't only working on machines understanding us, but also having them use voice as an identity.

"A person's voice is also his identity, just like his fingerprint. This is known as voice biometrics. Now Radha, who had to send money, won't need to remember a pin or password. She just has to say in her own voice, 'My voice is my identity.' And say, 'Send Rs. 500 to my daughter, Devi.' Her phone will recognize her voice and do this job for her."[117]

With Touch ID, Face ID from Apple voice only seems like the next logical step. It would open endless possibilities and ensure security in technology, so no one besides you could wire your money to someone else or even call someone else.

"Just think about it. When the process to recognize a voice, called 'speech recognition,' and the process to recognize a person through voice, called 'voice biometrics,' come together in our native language and dialect, we will have truly tamed technology forever.[118]

"With my innovation, whether you are a farmer or a CEO, whether you stay in a big city or a far flung region, whether you speak English or Bhojpuri or Tamil, your voice will have

---

117 Ibid.
118 Ibid.

a power that will help you embrace technology. Not just in India, but anywhere in the world."[119]

Combine this idea with the Amazon Alexa, Siri, and Google Voice and your possibilities are pretty much endless. Imagine when you said, "Alexa," it only answered to you, so no one else could exploit your Amazon Prime membership. I love that Sachev's focus is to make voice recognition and voice technology accessible to everyone.

JARVIS: Sir, Agent Coulson of SHIELD is on the line.

Tony Stark: *I'm not in. I'm actually out.*

JARVIS: Sir, I'm afraid he's insisting.

Tony Stark: *Grow a spine, JARVIS. I got a date.*

—*THE AVENGERS* (2012)[120]

119 Ibid.
120 "J.A.R.V.I.S./Quote."

CHAPTER 13

VOICE TECHNOLOGY CAN FURTHER EDUCATION

—

Pepper Potts: That's JARVIS. He runs the house.

—*IRON MAN* (2008)[121]

I look up to Noelle LaCharite greatly. She taught herself everything she knows. She was extremely passionate about her work for Alexa because she believes screens will become less and less popular and voice AI will increase. She has already seen it help education and the elderly.

---

121 Ibid.

Why type when you could just as easily speak to a device that would do the exact same thing for you. Google Homes, Amazon Alexa, and Siri are all steps in this direction. Having more voice technology is in the very near future and is a part of many young kids' lives already.

* * *

LaCharite started Class Alexa, which gives Alexa devices to classrooms, and she has also donated Alexa to retirement homes. The inspiration for these programs came from her family, and she is constantly thinking about how to better certain groups of people's lives through voice technology. Right now, 97 percent of classrooms have a computer in it for student use, and 58 percent of schools have laptop carts.[122] However, few schools have smart speakers in them.

"I recently had an epiphany about kids in homeless shelters or underprivileged children. You know my kids, we're obviously a bit more well off than the average, so my kids, of course, are getting exposed to the HoloLens, Mixed Reality and Alexa, but not everyone can afford an Alexa device. Some may not even know what it is. And I was like gosh we just have to get into these areas because it will literally shift the way people,

---

122 "Educational Technology in U.S. Public Schools: Fall 2018," *National Center for Education Statistics*, accessed January 10, 2019, https://nces.ed.gov/pubs2010/2010034.pdf

especially kids, think of what's possible, if they just know it's out there, and it's not fiction."

Tech like the Microsoft HoloLens and Mixed Reality have parallels in movies like *Star Trek* and *Iron Man* with the holograms you can see and touch in a 3-D manner. I used to think these weren't real, but the HoloLens makes these holograms appear right in front of you, so you can view files, emails, and other resources you need. Before researching all of this, it all seemed like fiction to me as well. LaCharite is trying to spread the word that this is a reality now, especially to underprivileged children.

"I am surprised when I go into a school, it depends on the school, but if I go into an inner-city school, none of them have seen an Alexa device unless maybe in the store like at Walmart or something. They don't own one, and they certainly don't understand the concept that they, even at their age of like nine or eleven, could build something for Alexa. They could go to library and do it for free in an hour.

"Now that I work for Microsoft, I've had exposure to the same thing being able to use images on Flickr or Google Photos or video from YouTube and be able to analyze it inside a browser. So again, these things are now accessible to anyone who can go to the library and get computer time. They could literally start a company and make money. It could change their lives, if they do, if they knew this was plausible."

* * *

The idea of getting technologies to underprivileged children is very important to me. As LaCharite mentioned, children with different problems in their lives will be able to think of how technology can solve these issues differently. Giving them access to computers would allow them to explore their interests and train them to use computers for their future jobs and work. It could also allow these children to create inventions that would solve major global issues.

As Raj Dhingra discusses in his TED Talk "Technology Will Help Educate Children for the Future," a student does not need expensive technologies to learn from computers. He brings up various examples of schools—from Turkey to California—of cheap computer technologies making a large impact in the classroom and giving the students the chance to learn at their own pace.

Both LaCharite and Dhingra are trying to show the world that you do not need a lot to make a large impact at a young age. Many students who could not afford an Alexa or an Apple computer have either cheap computer technology from their schools or public libraries available to them, but they do not know the lengths to which using these technologies could change their lives.

"They [Class Alexa] are focused on affluent communities, which is awesome. I think that's great. All kids should know, but they [affluent communities] are people who have already the easiest access to that technology. I just think ideas come from everywhere, and if we can give this to kids who have just a different perspective on life and have gone through harder times…

"You know, usually when you go through a hard time, like me with my dad, you find solutions for that space. So, imagine all these kids who have seen a hard life. They would have ideas about what they'd what to do to fix that and change that. And with cognitive services being able to use artificial intelligence, they could gather data. I just think there's huge opportunity."

LaCharite brings up a good point. Children who have seen larger problems could more easily come up with ways technology could solve these issues rather than just having Alexa set us a timer for our chocolate chip cookies to bake. This idea of the youth being more innovative and smart is seen in countless movies, books, and TV shows, such as *The Social Network* and *Good Will Hunting*.

I think the reason this kind of character is used infinitely many times is because it has a lot of truth to it. Once exposed to this technology, students who have faced issues like poverty, hunger, and serious crime in their young lives will be able to think of solutions much faster than those who haven't. As LaCharite said, when you face a certain kind of issue, you find a solution. But those of higher class face less problems and have less impactful solutions than those who have a different perspective and have gone through much harder times as a child.

Tony: Uh, say, JARVIS, is it that time?

JARVIS: The House Party Protocol, sir?

Tony: Correct.

—*IRON MAN 3* (2013)[123]

------

123 "J.A.R.V.I.S./Quote."

CHAPTER 14

VOICE TECHNOLOGY HELPS THE ELDERLY

Steve Rogers: *You remember that time we had to ride back from Rockaway Beach in the back of that freezer truck?*

Bucky Barnes: *Was that the time we used our train money to buy hot dogs?*

Steve Rogers: *You blew three bucks trying to win that stuffed bear for a redhead.*

Bucky Barnes: *What was her name again?*

Steve Rogers: *Dolores. You called her Dot.*

Bucky Barnes: *She's got to be a hundred years old by now.*

Steve Rogers: *So are we, pal.*

—CAPTAIN AMERICA: CIVIL WAR (2016)[124]

Noelle LaCharite still lives with her father. She believes the problems closest to you are the easiest to come up with solutions for, especially in technology.

Living with her father and being close to elder care has led her to think of ways to use technologies to support the elderly.

"I donated some Alexa to elder care. I always encourage people to go into assisted living facilities and just talk to people because they're just sitting in like this central area where everyone kind of congregates but they're not talking because they sit with the same people every day forever until they leave there.

"And I always say, 'Go in there and just have a conversation with them,' but wouldn't it be great if humans don't go in there, I think we should, but what if we don't? Wouldn't it be great if they could just have a conversation with a voice

---

124 "Winter Soldier/Quote," *Marvel Cinematic Universe Wiki*, accessed January 13, 2019, http://marvelcinematicuniverse.wikia.com/wiki/Winter_Soldier/Quote.

152 · 5 SCIENTISTS, 7 ENGINEERS, AND 2 AUTHORS MAKING YOUR SCIENCE FICTION DREAMS COME TRUE

service like Alexa and we have service-like personality chat where it's just like banter. It's just like 'hey, how's the weather?'"

* * *

LaCharite breaks off with a laugh during our conversation because her Alexa kept going off every time we would say her name.

"Oops, Alexa keeps saying hi to me every time I say her name, so I'm going to mute her. Oh, she just added something to my shopping list." She laughs. "So, I'll have to remember that."

* * *

"So, it's just great to give them, like my dad doesn't use it [Alexa] for much but he uses it every single day to find his favorite classical music station that he used to listen to as a kid or find his favorite books on audible and have them read to him as he's lounging in the backyard.

"Just making this accessible to an older community is almost the same for kids. They are enamored with the ability to not have to figure out how to look at an iPhone or a small smart screen. Even looking at screens like computers, it's easier for them to talk, right? And so me being able to roll back the technology of 'here's all the stuff you need to know in order to operate this website,' now you can just say what you want, and we call it top level intense. Whatever the top level thing you want to do let's make it easy for someone to ask for that.

"Again, and I find that every time I go in there, and we just have reading hour or we have Alexa read to whoever is sitting in that room, we have something to do and it's fun and Audible now is all about actors, so that's just one feature or something that's infinite in capability like Alexa or any of these speech services. So that alone just leads me to believe that there's so much more we could do to expose people to many not necessarily have access to these things.

* * *

"Like my grandmother, before she passed away last year, she had like nine brothers and sisters, but eventually she was the last one. She had her kids, but none of them lived in the state with her, so it's not like they'd come and just stop by. So literally she had no one, and I was just like man it'd be great to give people something like that. Something to talk to."

Countless times I have helped my grandparents set up their phones, look at photos posted online, or watch a YouTube clip; however, LaCharite explains that the elderly are a generation that could benefit greatly from voice technology.

I remember a time when my grandma was trying to look at photos from her grandson's first birthday, but she didn't know how to work the website. It was simply just typing a username and passcode, but she couldn't figure out how to reach the login page. Before talking to LaCharite, I used to just do the actions for them. Now, I have more of an incentive to teach them how to use the technology. It could help them a lot in the future—from looking to more photos and applying these skills to logging onto social media.

Right now, these kinds of voice technologies do not learn on their own, as LaCharite explained. Most of the conversation is just banter. However, people like LaCharite are working hard to develop ideas and technology that could learn as it goes rather than having a computer coded response to every question. I would have never thought of this as an area people in the tech world needed to focus on; however, because LaCharite is close to the issue, she is thinking about solutions that could help the loneliness for the elderly and increase their happiness, whether it be through conversation or listening to classical music.

Steve Rogers: I don't trust a guy without a dark side. Call me old-fashioned.

—*AVENGERS: AGE OF ULTRON* (2012)[125]

---

125 "Winter Soldier/Quote."

PART 5

LIVE ON THE CUTTING EDGE? NEVER GROW UP!

CHAPTER 15

MY MID-LIFE CRISIS

Bruce Banner: I don't know how to fly this thing!

Thor: You're a doctor. You have PhDs. You should figure it out.

Bruce Banner: None of them for flying alien spaceships!

—THOR: RAGNAROK (2017)[126]

If you're a geek, this will easily be your favorite part of this book.

Talking to Jim Kakalios, Becky Thompson, Sean Carroll, and learning about Adam Frank, brought out the fangirl in me. They all take their love of comic books and sci-fi movies

---

126 "Quotes From "Thor: Ragnarok,"" *Imdb*, accessed January 13, 2019, https://www.imdb.com/title/tt3501632/quotes/qt3703637.

to educate others. Kakalios, Carroll, and Frank have done consulting on movies, like *Thor* and *Doctor Strange* and are also all professors who educate college students.

Kakalios teaches a freshmen seminar class using his knowledge of physics and comic books.

"I read comic books as a kid. I gave them up in high school upon discovering girls."

Kakalios had a hilarious and geeky sense of humor that I loved. He knew exactly what interested me as a fangirl.

* * *

"In graduate school, I just happened to cross them [comic books] again, picked them up, and some of them were very good. Better than when I was a kid. And I just fell back into the hobby [reading comic books], basically as a diversion of my stress from my dissertation."

In 1988, Kakalios became a professor at the University of Minnesota. He taught classes, and at one point in the '90s, he tried to come up with an exam question with momentum and force that hadn't been done a million times already.

"It occurred to me that the death of Spider Man's girlfriend, Gwen Stacy, would be actually be a good illustration. Some students do a little sketch of the situation and identify the forces. Some students spent the entire exam period doing very detailed pencil innate drawings. The students seemed to enjoy taking a situation from a comic book and applying the physics principles.

"I tried to see if I could find other cases, and the thing that was striking about that is when you analyze the story line from a physics point of view, you find it is actually accurate."

* * *

Jim Kakalios is the author of the novel *The Physics of Super-heroes*. Kakalios's combination of his studies and superheroes started when he created the freshman seminar "Everything I Learned about Superheroes by Reading Comic Books," which ultimately turned into his book.

Kakalios started collecting examples of other stories in comic books that had accurate physics and showing these examples occasionally in his classes.

"You use the same joke right before you start to bring in some superhero. You say, 'For those of you wondering if there are any practical applications to the stuff you're learning...'"

Kakalios thought university freshmen seminars, which allow professors to come up with classes that aren't directly tied to the curriculum and are more interactive, would be a good opportunity to use more stories from comic books.

"Let me see if I can teach an entire class where the only examples came from superhero comic books. The class was a success; it went well, and in the spring of 2002, the first Spider Man movie was going to open. I wrote an essay for the *Minneapolis Star Tribune* about the death of Gwen Stacey and conservation and momentum.

"The *Spider Man* movie was a huge hit. The university put out a little press release about my class. Because of this harmonic conversion of two things, I wound up getting a lot of media attention, which is fun but also very scary. It's a thing where I don't know how my colleagues are going to take this, and I am trying to be engaging. And, I don't want people to think of me as a silly goofball.

"The one thing was very gratifying about it was that from all this media attention, I got hundreds of emails from students, teachers, and people who were long out of college, all really liking this idea of using superheroes to teach.

* * *

"This led to me writing my book, *The Physics of Superheroes.* It's called *The Physics of Superheroes*, but my working title when I was writing it was *My Mid-Life Crisis*. Even so, it turned out rather well."

Scott Lang: "I got something kinda big, but I can't hold it very long. On my signal, run like hell, and if I tear myself in half, don't come back for me."

Lang psyching himself up

"I'm the boss, I'm the boss, I'm the boss."

—*CAPTAIN AMERICA: CIVIL WAR (2016)*[127]

127 "The 17 Funniest Lines In the Marvel Cinematic Universe," *Screenrant*, accessed January 13, 2019, https://screenrant.com/ marvel-cinematic-universe-movies-funniest-lines-moments/.

CHAPTER 16

USING DISNEY CHANNEL AND COMIC BOOKS TO EDUCATE MIDDLE-SCHOOLERS

——

Samwell Tarly: I read it in a book.

—*GAME OF THRONES*[128]

I realized I may have never spoken to a more cheerful and excited person in my life. She laughs a lot, and put a smile on

---

128 "A Quote From A Game Of Thrones," *Goodreads.com*, accessed January 13, 2019, https://www.goodreads.com/quotes/227324-a-mind-needs-books-as-a-sword-needs-a.

my face as we were discussing her job. Most importantly, she is helping spread physics to younger students in an exciting way. I could tell that she understood the mindset of kids in school.

Becky Thompson received her PhD in physics at University of Texas Austin, but after a lot of days in the lab, she realized she wanted to do more public engagement. She went to the American Physical Society (APS) and decided to do public outreach.

"On the side I do some freelance work, writing comics for some who might be interested. I just submitted my book on the science of *Game of Thrones*. I did that through MIT press, which was a lot harder than I realized.

"It's interesting cause what I found out is that Game of Thrones is kind of as adult as adult gets. It's gory. I was researching 'here are fifteen ways to die,' or 'the science of incest.' It was very, very different than what I do day to day.

"At work, I manage a department, a lot of other people who are doing public outreach things, and then I write a series of comic books for middle schoolers about a superhero. They're full of terrible puns; just quite awful. They get to about half a million middle schoolers per year. I love that, and writing the *Games of Thrones* book, I was sort of like, 'You know what? I really like middle school kids better.'"

* * *

I have never met Thompson in person, but I could even tell from her lighthearted tone and constant laughter, writing educational comic books with corny jokes for middle schoolers could not have been a better job for her.

"I definitely would like to do another kind of pop science book because it was fun, but I have more of a middle school attitude. I am better at that.

"It was really interesting learning different sciences because I am a physicist by training, but a lot of the skills I learned doing research are things like how to read an academic paper or what questions to ask or how to find an academic paper.

"All the 'how to be a scientist' piece came hugely in handy when writing because I can figure out where to find biology papers, what they're saying, what graphs are important, where they are hedging on their results, and what their results are actually saying. That was really helpful in learning things like biology, which I don't know as much. That was weird, like biologists use terrible units, just really the strangest stuff, so that was interestingly complicated.

"In both things I do, it's a lot of pop culture. With the comics, it's not based on anything; I mean it's my own characters. It's a lot of like 'hey you like comics; let me use this character to teach you,' which is a different type of pop culture hook.

* * *

"In terms of how I ended up doing the [*Game of Thrones*] book at all, I knew someone who knew someone at MIT press, and I knew they were looking for pop science books. They were just like, 'Anyone want to write a book on something interesting?' So, I pitched this, and they said yes, and I ended up writing it. This was very much like, 'I don't know what I'm doing, but I have a great idea.' They said, 'Alright. We'll figure it out.'"

I was thinking as I was talking to Thompson, why would someone with such a love for kids and students write a book about the *Game of Thrones*? Most people have two sides to them, but the difference between middle school comic books and *Game of Throne*'s goriness is large.

"I love it [*Game of Thrones*]. I loved the book; I loved the show. A couple years ago, I was asked to do a dinner time lecture at a teacher's conference. It was for teachers in professional development. They asked me to train the teachers, which is great; I love doing that. I said, 'Well, what do you want me to talk about?' And they said they didn't care. Whatever I wanted. And I asked, 'Can you narrow that down? Are we talking about atomic physics or are we talking about anything I want?' And they said, 'I don't care whatever you want.' I was like alright, you asked, I mean your rules.

"I had just finished watching season 2 of *Game of Thrones*, and I thought here we go. I put together some of the science I had seen in the first two seasons, stuff about dragon fire and steel, in this talk. It went over really well, and people seemed really interested. I gave that talk at other places and that kind of morphed into a thirteen-chapter book from a couple of examples. A couple times I have been asked to give lectures to kids, like public pop science lectures. Obviously, I need to come up with something that doesn't involve beheading. I have other lectures on the physics of Wonder Woman and the physics on the movie *Frozen* just because I like those two. It's just a lot of what I like."

* * *

Thompson uses movies to connect people to science, which was my original inspiration for this book. Relating science back to something more common tends to intrigue people more. I really loved that Thompson had a similar philosophy, and it seemed to be successful for her with both children and adults.

"I love kids, particularly middle school. I don't think I ever matured past that age. I do well with middle school because it's different than being a teacher. I'm not trying to play an adult role. I'm just trying to have them have fun, so I am not trying to do anything structured. Not like, 'these are the learning goals,' so very different than a typical teacher interaction."

I think this point is key. A teacher is an authority figure, and I remember in elementary and middle school that no one liked their teachers. Thompson uses her comic books as a way to support learning and make the children intellectually curious. I think it's genius.

"Also, I still just think middle school jokes are funny. But, one thing I learned in grad school is that if you can't explain it to that level, you don't understand it. If you can't read this hugely complex thing and boil it down to three sentences that a middle schooler can get, you're missing something. They don't need you to go back and do all the differential equations and all the differential geometry. That's the last thing they need. What they need to know is this thing does this thing because of this thing."

I really liked this idea. I have many friends in higher level math or science classes than I am who cannot help me with my homework. It sounds weird, but it's true, and I think it relates back to Thompson's idea. If people cannot explain things on a basic level, they may not understand the main points of the lecture or lesson at all.

"I was talking to the medical advisor to the show *ER*, and he was saying you have three beats to get something across—three lines. So one person says this very complicated medical thing, and another person says this other very complicated medical thing, and the third person sums it up, saying, 'So if we don't do this extremely risky procedure, they're going to die.' You've got three sentences to get across the plot-driving point. I guess I kind of always used that since I heard that. Here is something borderline complicated, here is something that's maybe a reach, and the sum up is, if you do this, this happens. That gets it across to maybe the people who want the more complicated, but it also gets it across to others who just want the punchline."

If you watch almost any medical show like *Grey's Anatomy* or *ER*, you will never unsee this.

* * *

Thompson probably teaches better through her comic books than some teachers. I certainly believe that many teachers are caught up in explaining all the details that they forget to explain the basic concept, the punchline, at all. A lot of teachers, especially in the science world, expect kids to follow their intricate complex explanation well. I think Thompson breaks it down for the kids who want a more basic explanation.

"[The comic books are] mainly plot driven with science added in. Each one has four science activities that go with it. I design the activities first, and then kind of come up with a plot that ties them all together. I write the plot, and then make sure I am putting the activities in there.

"What is really interesting is having constraints makes it so much easier. If someone just said come up with a plot, at this point I have done enough where I could probably do it, but constraints spark ideas. I really like that. Say these are the four activities you need to get here and the overarching idea you're teaching is energy, and then you can kind of construct something.

"In terms of plot, another thing I do is, I love Disney Channel better than TeenNick. I watch a ton of Disney Channel to figure out what plots are working for them at that age. What are they interested in, what level are they emotionally, and how do they interact with each other and their parents? I take all of things into account when I write the comics for middle school. They are already watching these things, they are already drawn in by these things, these are already stories that are hooks. How can I use that to in my own story style?

"One thing that is interesting is that jokes we've heard a frickin million times that are completely trite, they've never heard before. This is the first time they're hearing it. I was watching a show a while back with the terrible joke, 'There's a fire at the circus. I heard it was intense.' At this point everybody has heard that joke, but this is the first time the middle schoolers are hearing it. There has to be a first time someone heard it. Stuff that we see as totally trite, terrible puns, all of this is hilarious to this group. A lot of it is researching what they'd like.

"The unfortunate part is now I am addicted. Now my TiVo is filled up with Disney Channel shows. I can pretend I'm watching them for work, but I really just like them."

Thompson clearly does her research on what kids love these days, and she is able to really connect with them like the Disney Channel does. She has educated the older generations that watch *Game of Thrones* to eleven-year-olds still watching Disney Channel. Her work with outreach will impact the next generation of scientists and is hopefully sparking students to go into STEM fields.

Tyrion Lannister: A mind needs books, as a sword needs a whetstone, if it is to keep its edge.

—*GAME OF THRONES*[129]

---

129 "A Quote From A Game Of Thrones."

CHAPTER 17.

FROM READING *THOR* COMIC BOOKS TO CONSULTING ON *THOR* FILMS

———

Thor: *I have no plans to die today.*

Heimdall: *None do.*

—*THOR* (2011)[130]

———

130   "Thor - Movie Quotes - Rotten Tomatoes," *Rottentomatoes.com*, accessed
      January 13, 2019, https://www.rottentomatoes.com/m/thor/quotes/.

When I thought of writing this book, Sean Carroll was the first person I wanted to talk to. He was also the person everyone else I interviewed told me to talk to. Carroll is an expert in the pop-science field and has consulted for movies on many different levels, ranging from student films to the Marvel Cinematic Universe.

His story is captivating; he went from a kid reading Thor comic books to being impactful when they became movies. To me, he seemed to be really living out his dream.

"I was one of those kids who just got lucky and fell in love with science when I was very young. Around ten years old, I would go to public libraries and read about science. I was in love with the idea of the black holes. That's what I wanted to do and fortunately, I was successful at doing it. I went to undergraduate school at Villanova and graduate school at Harvard, and then I did the usual thing of bouncing around from place to place until coming to Cal Tech in 2006."

* * *

Sean Carroll is a physicist, podcast host, and a verified Twitter user. When I was interviewing people to write this book, most people said Carroll is THE person to talk to about science in Hollywood.

Carroll went from reading comic books as a kid to consulting on Thor. Carroll's main job is being a physicist and professor at Cal Tech; however, he has dedicated some of his time working on blockbuster movies as well.

"The consulting [for sci-fi movies] is not a job at all. You don't get paid. You go in there for a couple hours, chat with people, and that's more or less the end of it. But I was happy to do it, and the Marvel comics were things I read as a kid growing up. A couple of my favorites were *Thor* and *Doctor Strange*, both of which were made into movies, so that made me very happy. I haven't been reading them for a long time, but I enjoy them, so I am glad to see them be so successful."

* * *

Adam Frank, who consulted on Doctor Strange, had a similar take on how to consult for Marvel movies and how to make an impact. Both Frank and Carroll have said that they help create the basis of the alternate world the movie is taking place in.

While working on a movie at Cal Tech, Carroll brought in some of his graduate students to help him. At first, they were confused about how to help; some of the movie plot was literally physically impossible. However, Carroll gave them another perspective of how to look at scripts and film production.

"The attitude I suggest is treat what happens in the script as data, as a theory. You don't have any choice about whether it's happened. It's happened and now you have to explain it. So once I put it that way, they really dug in. It is always possible to come up with something that makes for almost anything that could happen."

* * *

Carroll says that especially in Marvel movies, the timing of when the scientist comes in to help makes a large impact. The earlier in the script writing process they are, the easier it is to change, cut, and edit scenes. It makes sense; it's harder to change the script the closer to shooting.

"I should say it's very different to work on a Marvel movie then some other movies. The scientist can certainly have a bigger impact the earlier they come in. And, with Marvel movies, it's a machine. It's not even just one machine. There are many things going on with writers and directors and producers. It's going to happen. There are hundreds of millions of dollars at stake.

"The impact the scientist will have is relatively small. So what you can do is try to affect the spirit of how the science is portrayed, get a couple of pieces of language right or wrong, you can suggest some explanations, the fact that they Bifrost

Bridge is really a worm hole, but also, you can save them from mistakes.

"A scene that was supposed to happen in *Thor* where they were fighting on a planet, and it was a flat disk-like planet rather than a round planet. They fell off the edge, and we tried to say, you can't fall off the edge. Nothing is pulling you down. There is no gravity. So, they decided not to go with that in the final cut of the movie."

* * *

With Marvel, Carroll also says it's always plot and entertainment before science. It makes sense to me. These movies aren't really possible. I mean, I hope the *Infinity Gauntlet* isn't real. Tony Stark's technology is way too advanced to be a part of modern day, and there are no trees like Groot that can stretch their branches as far as they would like, or speak for that matter.

"None of the [Marvel] movies I have consulted on have tried to be strictly scientifically accurate. They're entertainment and stories first and they want science afterward, so it's not so much a matter of real world science as trying to make the world of the movie make sense and be logically coherent even if it's not the laws of physics we know and love."

"It's kind of fun to sit back and say well if there were advanced aliens or if you could upload your brain into a computer or something like that, how would that work if there was time travel? What rules would govern that? That kind of brain teaser of how things could work in such a world is the most exciting aspect."

* * *

Finally, I had to ask Carroll if he was friends with any of the actors from Marvel. Unfortunately, Carroll isn't friends with any of the Chrises (Evans, Hemsworth, or Pratt), but my face did light up when he told me he is friends with the actor who plays Agent Phil Coulson, Clark Gregg, who starts in Marvel's TV show *Agents of SHIELD* and some of the Marvel movies.

"Living in LA, it is almost hard to avoid [meeting people in the entertainment industry]. I've become good friends with Zach Stentz, who was a writer on *Thor*, the movie. And Clark Gregg, who plays agent Coulson in those movies. Other people you get to know by following on Twitter and things like that. The good ones are really creative, interesting, and intellectually curious. There are secret similarities between what scientists do for a living and what creative Hollywood people do for a living."

To me, Carroll has really come full circle. I mean he went from reading Thor comic books as a kid to actually impacting the movie, being friends with the writer and an actor in the Marvel franchise. Something I really admired about Carroll was that despite being a famous and well known scientist, he still dedicates his time to these movies. Yes, he has a personal connection to them, but he is also impacting the next generation of scientists through these movies. It's remarkable to think that someone at the highest level is still dedicating their time and not even receiving pay to help these movies, which have impacted my life immensely along with many others.

Thor: You are no match for the mighty Thor!

—*THOR* (2011)[131]

---

131 "Thor - Movie Quotes - Rotten Tomatoes,"

CHAPTER 18

HOW ADAM FRANK IMPACTED *DOCTOR STRANGE*

The Ancient One: Arrogance and fear still keep you from learning the simplest and most significant lesson of all.

Dr. Stephen Strange: Which is?

The Ancient One: It's not about you.

—*DOCTOR STRANGE* (2016)[132]

---

132 "Doctor Strange (2016)," *Imbd*, accessed January 13, 2019, https://www.imdb.com/title/tt1211837/characters/nm0842770.

Doctor Stephen Strange is a neurosurgeon gone magical sorcerer in the Marvel Cinematic Universe's (MCU) Doctor Strange. Doctor Strange is very focused on reality, and Adam Frank, the scientific consultant for the movie, took on the role of advising the movie on illustrating the human experience of space and time.[133]

Frank says MCU has respected science throughout most of its films and set some rules the movies follow.

"Compared to [the DC movies], they've imagined there's a science at work. You know, Tony Stark and what he can do with the devices he builds and the idea that Thor, the inhabitants of Asgard, are actually aliens, magic … a number of times they've used that Arthur C. Clarke quote: 'Any advance of technology looks like magic to other people.'"[134]

Frank goes on to say MCU has incorporated science in their past films well and they balance their plots with science well.

"In the *Thor* movies, there's the Einstein-Rosen Bridge, so the rainbow way or whatever it's called is actually a wormhole. In *Ant-Man*, they brought in quantum mechanics. And the

---

133 Neel V. Patel, "A Physicist Explains Consciousness to 'Doctor Strange,'" *Inverse*, accessed January 13, 2019, https://www.inverse.com/article/21937-doctor-strange-consultant-adam-frank.

134 "A Physicist Explains Consciousness to 'Doctor Strange,'"

great thing about Marvel is they use those devices as twists in the story to make the story move along."[135]

* * *

With *Doctor Strange*, it was probably more complicated than Falcon or Black Widow because Strange has powers that aren't physically possible, but Frank had to find a way to set some ground rules of the impossible.

"The dilemma with *Doctor Strange*, to me, was to find a way to embrace that vision and that respect for science for a character who—in the comic books at least—is a magician, an occult master. And so how do you get that into the Marvel Universe? [The film's producers] wanted to respect everything they've been slowly and carefully building."[136]

Frank was contemplating how to fit Strange into MCU. His original thinking was to use neurological forces; however, he diverted from that idea and focused more on the idea of consciousness.

"We don't really have a handle on what consciousness is. We don't have a scientific, materialist, reductionist account

---

135 Ibid.
136 Ibid.

for consciousness yet. Maybe we will have one, but it's also entirely possible that we will not have one and we will need to add other things to have a proper science of consciousness.[137]

"A noted philosopher, David Chalmers, came up with the idea of the "hard problem" in the 1990s. He was trying to explain how neuroscience has made interesting progress on things like vision—how does vision actually work, how do you represent something from visual input, which part of the brain is active? For him, the hard problem was how to account for the personal vividness of experience. Even if I wrote down an equation for you eating an apple, that would be very different from your experience of eating the apple."[138]

* * *

Frank then goes on to discuss the idea of the multiverse: something he doesn't believe is true but worked on in the movie. The multiverse is the idea that, after the Big Bang, multiple universes were created, not just one. Each universe is separate with a different set of physical laws.

"That idea comes from cosmology, but physicists have been playing around with lots of different ideas of multiverses. For

---

137 Ibid.
138 Ibid.

example, in quantum mechanics (and trying to understand the weirdnesses of quantum mechanics), people have proposed that every time a quantum event occurs, the universe splits off into a bunch of parallel dimensions, each with a different quantum result, and then each of those individual universes goes on and evolves on its own. And then, of course, there'll be more quantum events and then there'll be more splittings until you have an infinite number of parallel universes.[139]

"What you could do in a movie is imagine that you're getting access between all of these universes, like the universe that we're living in is one universe out of an infinite number of universes in the multiverse. So the multiverse is all the possible universes, and all these different universes have utterly different kinds of laws, different behavior, so you can use that idea in a story to allow your characters—especially as they're using consciousness, using this openness of consciousness in the story—to be able to have access and move between these different universes.[140]

"In the *Doctor Strange* comics, he's always going to these different dimensions. And that was very 1960s, totally groove-orific, trippy. So a way of bringing that into modern physics

---

139 Ibid.
140 Ibid.

parlance is [thinking about how the] different dimensions become the different universes in the multiverse."[141]

* * *

In *Doctor Strange*, Strange is taught by the Ancient One, and Frank said he helped create the dialogue between these two characters: Strange, a doctor and realist at the beginning of the movie, and the Ancient One, a believer of faith and what most think is impossible.

"What would it look like between a materialist, rationalist, reductionist and someone who had this enlarged perspective. Because I'm a physicist, I know exactly what those materialist arguments look like; I've also written on science and human spirituality. I'm not a reductionist. I'm an atheist, but I'm not a reductionist. So that was one of the things we sort of went back and forth with.[142]

"And that was fun, just talking about what somebody who believes there was nothing to consciousness except your neurons would argue with someone who was saying, 'No, there's actually more.' What would that look like?" That was fun and super awesome for me. I'm looking forward to seeing

_____

141 Ibid.
142 Ibid.

the final project, but what I've seen so far I've been really excited about."[143]

I think that this concept is a very interesting idea. There are a lot of people who believe there is God or something else outside of science. Although I am non-religious, I have many friends who are religious, and what Frank is talking about is not dissimilar to conversations many people have fairly often. These are not concepts I would have thought went into the script-writing process of a movie.

The Ancient One: We never lose our demons, Mordo. We only learn to live above them.

—DOCTOR STRANGE (2016)[144]

143 Ibid.

144 John McAteer, ""We Never Lose Our Demons, We Only Learn To Live Above Them,"" *Video Ut Intellectum*, accessed January 13, 2019, https://filmphilosopher.wordpress.com/2016/11/06/we-never-lose-our-demons-we-only-learn-to-live-above-them/.

CHAPTER 19

LIFE AT COMIC CON

———

Uncle Ben: With great power there must also come great responsibility.

—*AMAZING FANTASY #15*[145]

I read a lot about Thompson's work with comic books for middle schoolers, but I also saw she had attended San Diego Comic Con (SDCC). It has been a dream of mine for as long as I can remember to go, get all dressed up, and nerd out with everyone else there. I always imagined the long lines waiting for panels and seeing some of my favorite sci-fi actors. Inevitably, I had to ask what this insane event was like.

———

145 "With Great Power Comes Great Responsibility – Quote Investigator," *Quoteinvestigator.com*, accessed January 15, 2019, https://quoteinvestigator.com/2015/07/23/great-power/.

"I've been there [San Diego Comic Con] nine times. We go to that, we go to Awesome Con in DC and possibly Denver Comic Con, which is very education based, But San Diego Comic Con, I have never in my life seen anything like it, like ever. There is absolutely no amount of explanation that can accurately describe the insanity that is San Diego.

"I mean we've done it for a long time, and a couple of us have been going for a couple of years, and so it's bad enough we know comic con hacks of like how to get places. It is exceptionally difficult to walk down the street at certain times. At certain hours people are going in one direction, and like if you want to go back the other, you can't, but there's like side streets you can go around. We duck in one specific door so we can avoid all the crowds trying to get into the exhibit hall and it's just things like that. It's insane.

"What you have to plan on is crazy, and we usually stock up on food from the grocery store because you can't buy food. It's crazy. They run out of food. So you end up with this problem where you have an hour off the booth because we cycle through booth time. You can't walk far away because off the traffic patterns, you can't get food because either the lines are too long or they're out. It's crazy, so for a week, you live off hot dogs from the food court down the street or Mrs. Field's cookies."

* * *

As she was explaining all of this, I imagined myself, all dressed up, probably as Padme Amidala or Scarlett Witch, and I needed to know if she and her coworkers had seen any panels. Panels at SDCC are when the casts from huge blockbusters like *Star Wars* or hit TV shows like the *Flash* come out and answer questions. They also usually drop news about their next season or part to their film franchise.

"I do the booth schedule, so five of us [from APS] go, and I ask everyone to send me three panels they really want to go to. I make sure to give them time off for those three panels, and then do the booth schedule around that.

"I really want to go to Wonder Woman this year, and I am actually emailing the press person saying, 'Hi, I do a talk on the physics of Wonder Woman, get me in!' It won't work, but it's worth a try. So we do that, but the thing is the good panels, like the DC panel, the Marvel panel, there is one big room called Hall H, and you have to sleep out on the sidewalk to get into Hall H. You usually have to wait in line for twenty-four hours to get into the panel you want."

In the past Hall H panels have included Teen Wolf, the Big Bang Theory, Marvel's the Defenders, and Stranger Things.

* * *

"It's set up to do it. They have the lines set up with tent covers, space for your air mattress, and you're going to sleep in line. They issue wristbands in the middle of the night, so people with wristbands know they can go to the bathroom if they need to and get back in line. People are usually pretty respectful. Domino's Pizza delivers to the Hall H line. It's crazy!

The Ballroom 20 line is also insane, so you know which ones you're not going to be able to get into. A couple of years ago, I really wanted to see the *Torchwood* panel in Ballroom 20, but that day they were up against some really good stuff in Hall H, so the hardcore people that were going to sleep out probably were going to sleep out all night for Hall H, but not Ballroom 20.

"I got up at 4 a.m., went to the Ballroom 20 line, slept on the pavement until they started moving us at 9 and then got to go see the Torchwood panel. I did the X-Files panel the same way. It was worth it. No question."

As Thompson was explaining all of this, my inner fangirl was coming out, imagining myself potentially being in the same room as Chris Evans or Sebastian Stan. Very few people can relate to the fact that it was worth it to wake up at 4 a.m. and sleep on the pavement to see a comic con panel. Most people think it's outrageous. I respect it.

* * *

Thompson goes for work to promote the American Physical Society and its programs. She is clearly also a fan, but something I respect from her booth is that it is there to educate the youth that go to SDCC about science fields.

As a woman interested into going into technology or STEM-related business, this made a large impact on me and made me realize how important her work in science outreach is at the American Physical Society. Although there are many actors and people dressed as heroes at SDCC, Thompson and her team make a real impact, especially on those of a younger generation.

C-3PO: Don't call me a mindless philosopher, you overweight glob of grease!

—*STAR WARS EPISODE IV: A NEW HOPE*[146]

146 "6 of C-3PO's Best Insults | Starwars.com," *Starwars.com*, accessed January 13, 2019, https://www.starwars.com/news/6-of-c-3pos-best-insults.

PART 6

THE FUTURE IS NOW AND OURS

CHAPTER 20

SPACE WILL AFFECT LIFE ON EARTH

Dr. Floyd: [prerecorded message speaking through TV on board Discovery while Bowman looks on] Good day, gentlemen. This is a prerecorded briefing made prior to your departure and which for security reasons of the highest importance has been known on board during the mission only by your H-A-L 9000 computer. Now that you are in Jupiter's space and the entire crew is revived, it can be told to you. Eighteen months ago the first evidence of intelligent life off the Earth was discovered. It was buried forty feet below the lunar surface near the crater Tycho. Except for a single very powerful radio emission aimed at Jupiter the four-million-year-old

black monolith has remained completely inert. Its origin and
purpose are still a total mystery.

—*2001: A SPACE ODYSSEY* (1968)[147]

"I was seventeen when I chose my career," Danielle Wood says as she opens her TED Talk in 2017. "I was standing outside on a hot summer night in Florida, just a few miles from the ocean. I was waiting for a miracle to happen.[148]

"That summer, I was privileged to work as an intern at NASA's Kennedy Space Center, and the miracle I was waiting for was the launch of the Columbia Space Shuttle carrying the Chandra X-Ray Observatory, a telescope that would allow scientists to peer into the edge of black holes. The entire sky filled with light. And it was as if it was daytime in the middle of the night. Soon, we could feel the rumble of the engines vibrating in our chests. And it wasn't a miracle; it was the combined effort of a team of thousands of people who worked together to make what seemed impossible a reality. And I wanted to join that team."[149]

---

147 "Quotes From "2001: A Space Odyssey,"" *Imdb*, accessed January 13, 2019, https://www.imdb.com/title/tt0062622/quotes/qt0396927.

148 Danielle Wood, "6 space technologies we can use to improve life on earth," posted 2017, accessed January 13, 2019, mp4 format, https://www.ted.com/talks/danielle_wood_how_we_can_use_space_technology_to_improve_life_on_earth?language=en.

149 Wood, "6 space technologies we can use to improve life on earth."

Wood is talking about the driving force behind my book—making what seemed impossible a reality. What Wood describes is something most have not experienced before—technology opening up new avenues for exploration. Wood's work really explores the endless possibilities of space to assist humans on Earth.

* * *

Wood went to MIT where she was able to study aerospace engineering. She started to learn engineering training and how to make space robots, but when she traveled to Africa, she realized she wanted to be a part of a group that supported women.

"Now, my confusion arose in my summer breaks. I traveled to a school in Kenya, and there I volunteered with girls ages five to seventeen, giving them lessons in English, math and science. They taught me songs in Swahili. And mostly, I just spent time getting to know the girls, enjoying their presence. And I saw that these girls and the leaders in their community were overcoming important barriers to allow these girls to have the best possible chances in life. And I wanted to join that team."[150]

---

150 Ibid.

Wood wanted to improve girls' lives around the world. She feared her chosen career wouldn't be able to help girls everywhere; however, she was mistaken.

* * *

Wood went back to intern at NASA where she learned that countries like Kenya had been using space technology to help themselves and improve the quality of life there.

When at NASA, she learned she could have a career in both space and in development.

"This idea is not new. In fact, in 1967, the nations of the world came together to write the Outer Space Treaty. This treaty made a bold statement, saying, 'The exploration and use of outer space should be carried on for the benefit of all peoples, irrespective of their level of economic or scientific development.'[151]

"We have not truly lived up to this ideal, although people have worked for decades to make this a reality. Forces such as colonialism, racism, and gender inequality have actually excluded many people from the benefits of space and caused us to believe that space is for the few, the rich or the elite. But

---

151 Ibid.

we cannot afford this attitude because the world is engaged in a vital mission to improve life for everyone."[152]

* * *

When I think of space, I used to think of just a few astronauts going to space, discovering something or experimenting. Then, they return to Earth as a small group of heroes. This can be seen in a variety of space movies, like *The Martian*.

However, Wood has been working to make the greatness of space accessible to everyone, and she is trying to uphold the ideas of the Outer Space Treaty.

"Our road map for this mission comes from the seventeen Sustainable Development Goals of the United Nations. All the member states of the United Nations have agreed that these are the priorities between now and 2030. These goals give us our key moments and opportunities of our time—opportunities to end extreme poverty and ensure everyone has access to food and clean water. We must pursue these goals as a global community, and technology from space supports sustainable development. In fact, there are six space services that can help us pursue the Sustainable Development Goals."[153]

---

152 Ibid.
153 Ibid.

Space is not something I thought about being helpful to large world problems, such as poverty and clean water. I always imagined space as those insane star-exploding images people have as their iPhone wallpapers. In fact, I had no idea the UN had any sort of initiative to use space for the greater good.

Wood goes on to name technologies that will make the Earth a better place to live and increase the quality of life here.

"Communication satellites provide access to phone and internet service to almost any location on Earth. This is particularly important during times of disaster recovery. When Typhoon Haiyan struck the Philippines, the local communication networks needed to be repaired, and teams brought in inflatable communication antennas that could link to satellites. This was useful during the time of repair and recovery."[154]

"Positioning satellites tell us where we are by telling us where they are. Scientists can use this technology to track endangered wildlife. This turtle has been fitted with a system that allows it to receive location information from positioning satellites, and they send the location information to scientists via communication satellites. Scientists can use this

---

154 Ibid.

knowledge to then make better policies and help determine how to keep these animals alive."[155]

"Earth observation satellites tell us what's going on in our environment. Right now, there are about 150 satellites operated by over sixty government agencies, and these are just those observing the Earth. Meanwhile, companies are adding to this list. Most of the governments provide the data from the satellites for free online."[156]

"In space, we have an orbiting laboratory on the International Space Station. The vehicle and everything inside are in a form of free fall around the Earth, and they don't experience the effect of gravity. And because of this, we call it 'microgravity.'"[157]

"When astronauts are in the microgravity environment, their bodies react as if they're aging rapidly. Their bones and muscles weaken, and their cardiovascular system and their immune system change. As scientists study how to keep astronauts healthy in space, we can take the exercises and techniques we use for astronauts and transfer them to people on Earth to improve our health here.[158]

---

155 Ibid.
156 Ibid.
157 Ibid.
158 Ibid.

"Often, as we develop technology for astronauts and exploration or for spacecraft, we can also transfer those inventions to improve life on Earth. One of my favorites is a water filtration system. A key component of it is based on the technology to filter wastewater on the space station. It's now being used around the world.[159]

"Space is also an infinite source of inspiration, through education, through research and astronomy and that age-old experience of stargazing. Now, countries around the world are engaging in advancing their own development by increasing their local knowledge of engineering and science and space."[160]

Wood is not only using her work to better her communities on Earth, but her goal is to make these benefits accessible to everyone. From water filtration systems to tracking endangered species, these technologies could be useful all over the globe. Wood and the UN are trying to make that possible.

* * *

She ends her speech by showing us young aerospace engineers from all over the globe including a girl from Venezuela and a boy from the Philippines. Now, all countries and many

---

159 Ibid.
160 Ibid.

people around the world have heard about these beneficial space technologies to help support their countries.

Although it may seem like we are developing futuristic ideas, Wood says there is still so much more work to be done.

"We have more work to do because there are still barriers that exclude people from space and limit the impact of this technology. For many people, Earth observation data is complex, and satellite communication services are too expensive. Microgravity research just appears to be inaccessible. This is what motivates my work as a professor at MIT's Media Lab.[161]

"I've recently founded a new research group called Space Enabled. We are working to tear down these barriers that limit the benefits of space. And we're also going to develop the future applications that will continue to contribute to sustainable development. We'll keep on this work until we can truly say that space is for the benefit of all peoples, and we are all space enabled."[162]

Similar to many of the people featured in part 3 (AI), Wood really wants to use the power of technology to better human

---

161 Ibid.
162 Ibid.

life. Wood is another person who convinces me that technology will only help us and not hurt us.

I loved Wood's determination and drive to continue to make technology more accessible. A lot of the people I have been talking with want to help create technologies that make life easier and make them accessible to everyone. Whether it's education, quality of daily life, or the elderly, technology needs to be spread and can make a large impact on a large scale. Wood is one of many leaders who are using technology to better life.

The Book: *Space, says the introduction to the guide, is big, really big. You just won't believe how vastly, hugely, mind bogglingly big it is. And so on.*
—*THE HITCH HIKER'S GUIDE TO THE GALAXY* (2005)[163]

---

163 "Awe And Amazement Quotes: The Hitchhiker's Guide To The Galaxy Page 2," *Shmoop.com*, accessed January 13, 2019, https://www.shmoop.com/hitchhikers-guide-to-the-galaxy/awe-and-amazement-quotes-2.html.

CHAPTER 21

TECHNOLOGY CAN REDUCE URBANIZATION

———

Thrawn: What about Coruscant? Is there unrest here?

—STAR WARS: THRAWN[164]

Imagine a planet that is full of skyscrapers and lights all the time. I imagine Coruscant, the planet that is just one big city that makes many appearances in the prequel trilogy of *Star Wars*.

———

164 "Coruscant," *Wookieepedia*, accessed January 13, 2019, http://starwars.wikia.com/wiki/Coruscant.

Coruscant was the capital of the Republic and the birthplace of human life. It was the planet that never slept. The majority of the world population lives in cities, and in 2050, supposedly 66 percent of the population will live in cities.

Organizations like the United Nations and the World Health Organization are telling people we need to plan for overpopulation and plan for more sustainable cities.

* * *

Despite the trends, Julio Gil believes that urbanization will decline as time progresses. He believes the future of technology will give you the benefits of city life in rural areas. According to Gil, people move to cities for three main reasons: more job opportunities, easier access to services and goods, and a rich social life.[165]

Gil has worked for UPS Corporate for fifteen years driving innovation and logistical improvements.[166] He is creating new technologies for drones, 3-D printing, and more.

165 Julio Gil, "Future tech will give you the benefits of city life anywhere," posted 2017, accessed January 13, 2019, mp4 format, https://www.ted.com/talks/julio_gil_future_tech_will_give_you_the_benefits_of_city_life_anywhere?language=en
166 Gil, "Future tech will give you the benefits of city life anywhere."

More than 80 percent of people say they would rather work from home according to Global Workplace Analytics.[167]

It costs a company $11,000 a year per employee to have an office. If half those workers worked electronically or from home, the savings in the US would exceed 500 billion dollars.[168] Gil explains that although teleworking can seem ideal, current technology makes it feel isolating.

* * *

During a TED Talk called "Future tech will give you the benefits of city life anywhere," Gil explains how Augmented Reality has already changed the workplace and will continue to do so.

"Augmented reality already today allows you to take your office environment everywhere with you. All you need is a wearable computer and a pair of smart glasses, and you can take your emails and your spreadsheets with you wherever you go. Video conferences and video calls have become very common these days, but they still need improvement. I mean, all those little faces on a flat screen, sometimes you don't even know who is talking."[169]

---

167 Ibid.
168 Ibid.
169 Ibid.

When I heard Gil say this, I could easily relate. I study Russian, and my class has been the same students for the entire four years of high school. Before every test, we would all video chat on an app called House Party. The app was great at first because we weren't used to having so many people video chatting at once. FaceTime doesn't allow multiple people to be on a call at once. However, as our study sessions continued, less people would join because it was just too many people's tiny faces on a screen at once. It became ineffective. There was so much audio overlap, and we would have no idea what everyone else was saying. Gil provides a solution for this problem.

He goes on to discuss how "tablet on a stick" could revolutionize the workplace and give people mobility in the office even when they aren't physically there.

"Now, we already have something way better than static videocalls—your average telepresence robot. I call it tablet on a stick. You can move around, and you can control what you're looking at. It's way better, but far from perfect. You know how they say that most human communication is nonverbal? Well, the robot doesn't give you any of that. It looks like an alien. But with advances in augmented reality, it will be easy to wrap the robot in a nice hologram that actually looks and moves like a person. That will do it. Or else, forget the robot. We go full VR, and everybody meets in cyberspace. Give it a couple of years and that will feel so real, you won't tell the difference."[170]

* * *

Ultimately, Gil is saying that technology will advance enough to give people what they want—to work from home. It will also give people the feeling of collaboration and being physically there without having to leave your bedroom.

This means that to work, one doesn't have to live even remotely close to their office building; they just need the right technology to feel like they are in their office. However, I do feel like these ideas are very, very futuristic. In part 1 of this book, Professor Curtis Broadbent explained how it will take a very long time to have holograms (3-D volumetric displays) in free space. There is still no technology that does that safely. I think it is possible but only in the distant future.

* * *

After a long day of school or work, no one wants to go home and take the time to cook their dinner for one. This is why GrubHub, Postmates, and Door Dash exist. These food services can instantly bring you your favorite burrito from

---

170 Ibid.

Chipotle or your custom salad from Sweetgreen by simply pressing a button. It's almost too easy.

The second reason Gil thinks people move to a city is services and goods. Online shoppers already take up a significant portion of retail. Stores are closing everywhere. However, e-commerce to rural areas takes forever, and they are the most expensive. Gil talks about an easy and already existing fix to this problem—drones.

"A vehicle carrying a squadron of drones. The driver does some of the deliveries while the drones are flying back and forth from the truck as it moves. That way, the average cost for delivery is reduced, and voila: affordable e-commerce services in the countryside. You will see: the new homes of our teleworkers will probably have a drone pod in the yard. So once the final mile delivery is not a problem, you don't need to be in the city to buy things anymore. So that's two."[171]

With the help of drones, everything can be delivered anywhere. Gil makes people think about this question: Why live in a city when everything you need can come right to you any time of day?

* * *

171 Ibid.

As soon as I stop hanging out with my friends or go home after school, I check my text messages, social media accounts, and email. People want to make friends. They want to chat, gossip, and flirt. People think they need to be in the city for this, but a lot of people already do this from the comfort of their own sofa.

There are billions of social media users in the world, which makes people think we are connected no matter where we are. This is true, but sometimes people need real human interaction as well. Gil explains why the city is not necessarily the best place for human interaction, how moving to a rural area can actually make small businesses thrive, and why moving to rural areas can make us more eco-friendly.

"Sometimes you still need some real human contact. Ironically, the city, with its population density, is not always the best for that. Actually, as social groups become smaller, they grow stronger. A recent UK study by the Office for National Statistics showed a higher life satisfaction rating among people living in rural areas.[172]

"As people settle in the countryside, well, they will buy local groceries, fresh groceries, foodstuff, maintenance services. So handymen, small workshops, service companies will

---

172 Ibid.

thrive. Maybe some of the industrial workers from the cities displaced by the automation will find a nice alternative job here, and they will move too."[173]

Gil brings up a point that I have not yet addressed—automation displacing workers. There are long debates about whether automation of lower and middle class jobs is ethically okay. As David Lee explains in part 3 of my book, I do think it will open up more people to innovative jobs. According to inc.com 55 percent of Gen Z want to start their own business.[174] However, people still want the freedom of choice. It's hard to know the solution to the problem. Companies keep using automation because it is cheaper and makes products faster.

"And as people move to the countryside, how is that going to be? Think about autonomous, off-the-grid houses with solar panels, with wind turbines and waste recycling utilities. Our new homes will produce their own energy and use it to also power the family car. I mean, cities have always been regarded as being more energy-efficient, but let me tell you, repopulating the countryside can be eco too."[175]

---

173 Ibid.
174 Gherini Anne, "Gen-Z Is About to Outnumber Millennials. Here's How That Will Affect the Business World," *Inc.*, accessed January 10, 2019, https://www.inc.com/anne-gherini/gen-z-is-about-to-outnumber-millennials-heres-how-that-will-affect-business-world.html
175 Ibid.

* * *

Gil and his family made the move themselves, and they have never looked back.

"Six years ago, my wife and I packed our stuff, sold our little apartment in Spain, and for the same money we bought a house with a garden and little birds that come singing in the morning. It's so nice there. We live in a small village, not really the countryside yet. That is going to be my next move—a refurbished farmhouse, not too far from a city, not too close. And now we'll make sure to have a good spot for drones to land."[176]

Although Gil really vouches for rural life, he knows that urban life will not end. If you don't like the countryside don't come because it is not place for you. He is telling those who would rather live in the country side that it has the same benefits of living in a city.

A lot of data says that urbanization isn't ending, though. Since 1950, 30 percent more of the world's population has moved to cities.

---

176 Ibid.

There are twenty-nine megacities with more than ten million people living in them. People love their cities, especially in North America where 82 percent of the population lives in a city. These all seem like counterarguments to Gil's point; however, I think Gil's main concept was that if anyone feels like they "have to be" in a city, they don't.

City life, the hustle and bustle, is cut out for some people, but others would rather live in rural areas. They feel as if their jobs or social life are keeping them in a megacity like New York or Los Angeles.

Gil is saying these people have the option to leave, and people love having options and being able to pick their destiny. The Countryside is not for me, but it is great to know I have the option.

Admiral Niles Tenant: *Only some secrets are well kept on Coruscant.*

—*STAR WARS: TARKIN*[177]

---

177 "Coruscant"

CHAPTER 22

BLACK PANTHER'S LETITIA WRIGHT ENCOURAGES YOUNG GIRLS TO GO INTO STEM FIELDS

———

Martin Freeman's character Everett Kenneth Ross wakes up in Wakanda.

Everett Kenneth Ross: Is this Wakanda?

Shuri: No, it's Kansas.

—*BLACK PANTHER* (2018)[178]

Shuri, played by Letitia Wright, is the young female scientist, and the Black Panther's younger sister in Marvel's blockbuster hit and cultural phenomenon *Black Panther.*

Shuri is portrayed to be smart, a little bit of a wise-ass, and holds her own throughout the movie. Although Shuri is only sixteen, she is one of the smartest characters in the whole Marvel Cinematic Universe.

After playing the role of Shuri, Letitia Wright teamed up with Shell to inspire girls to go into STEM (Science, Technology, Engineering, and Math) fields.[179] Wright is creating a video called Engineering Real-Life Heroes, which is about four females in STEM. The video profiles each of these four girls from diverse backgrounds and their plans to become STEM leaders. Wright hopes Shuri will make an impact on a younger audience.

---

178 Hunter Harris, "Here Are Shuri'S Best Black Panther One-Liners," *Vulture*, accessed January 13, 2019, https://www.vulture.com/2018/02/black-panther-quotes-shuri-one-liners.html.

179 "Engineering Real-Life Heroes," *Shell*, accessed January 13, 2019, https://www.shell.co.uk/energy-and-innovation/make-the-future/engineering-real-life-heroes.html

"I hope it inspires them and I hope it does [so] positively. I hope it sparks the next person."[180]

The film brings together Wright along with Anne-Marie Imafidon, the creator of Stemettes, and they show these four girls preparing for the Shell-Eco marathon—a competition to see who can create a fuel-efficient vehicle.[181]

* * *

Wright has spoken out often about this project and STEM, and she wants this video to help young girls start to believe these are possible career paths for them.

"I strongly believe you have to see something in order for you to understand you can do it. That's why I'm thrilled that I got to tell the stories of these incredibly talented young women, who are real-life embodiments of what STEM really is about. It would mean a lot to me if a girl could watch this film and think they too could do what these young women do."[182]

She also says many people have told her that her representation of a female engineer and scientist means a lot to them.

---

180 "Engineering Real-Life Heroes,"
181 Ibid.
182 Ibid.

"I have gotten a flood of messages from people saying thank you for representing us. There's been such a lack of exposure to young women in STEM subjects."[183]

* * *

Wright has always been on the lookout for roles that would make a difference and show underrepresented races or types of people.

"Since I was seventeen, I've tried to break the mold of what we see on TV and film, especially as a young, black woman," Wright said. "I want to see something different. You have to ask the question—why not see me for this? Why not see me for this different role, this quirky character?"[184]

The video went live on Shell's YouTube page on June 25, 2018.

Black Panther was one of the most inspiring movies of the decade. It had great representation, and it feature empowered female characters like Shuri and Nakia.

There are large debates about whether Shuri is the smartest person in the MCU. (I personally think she and Stark are tied in terms of intelligence, but she's a teenager, so ultimately she wins.)

---

183 Ibid.
184 Ibid.

* * *

Black Panther was part of the reason I decided to write this book; I was so utterly inspired by Shuri's technology. Wright has easily inspired me alongside many other young women to create and take action in STEM fields. According to the National Girls Collaborative project, all STEM fields are still less than 50 percent women;[185] however, there has been a 27 percent increase in female leadership hires in Software and IT services.[186] Advocates like Wright are starting to make an impact.

Shuri: This corset is really uncomfortable, so can we all wrap this up and go home?

—*BLACK PANTHER* (2018)[187]

* * *

Wood, Gil, and Wright are all using their power to change the future through STEM. Wood wants to show the world that space can be used to change everyone's lives and make

185 "Statistics," *National Girls Collaborative*, accessed January 10, 2019, https://ngcproject.org/statistics

186 Mejia Zameena, "More women entered STEM over the past 40 years than any other field, new data shows," *CNBC*, accessed January 10, 2019, https://www.cnbc.com/2018/03/07/more-women-entered-stem-over-the-past-40-years-than-any-other-field.html.

187 "Here Are Shuri'S Best Black Panther One-Liners,"

the Earth a more sustainable place. Through technology, Gil explains how people who live in urban areas but want rural life can have it. Wright is using her influence as an actress to help young women go into a field that is mainly male dominated. They are using STEM to change the future and make the impossible possible.

CONCLUSION

Although there is still a lot of work to be done, I, along with many of the people in this book, believe that technology will brighten our future. Even though people think technology could take over the human race, many people are ensuring that does not happen and attempting to prevent every bad outcome.

Today, I hear more people discussing the negative qualities of technology rather than the upsides. I hope this book has helped you realize that technology will assist humans, and a variety of new technologies are on the rise that would revolutionize health, the workplace, and other aspects of everyday life.

Obi-Wan Kenobi: Remember, the force will be with you, always.

—*STAR WARS EPISODE IV: A NEW HOPE* (1977)[188]

---

188 "View Quote ... Star Wars Episode IV: A New Hope ... Movie Quotes Database."

ACKNOWLEDGMENTS

———

First and foremost, I'd like to thank my parents for giving me this opportunity. Mom and Dad, thank you for your constant support throughout the writing process. Mom, thank you for always helping me plan my time with my busy schedule and teaching me how to destress when my worries get the best of me. Dad, thank you for always pushing me to achieve the best.

Thank you to all my interviewees. Thank you for taking the time to talk to me about your exceptional work. This book would be nothing without you.

Thank you to my high school friends, Sophie and Halley, who took the time to read this book thoroughly and give me fantastic feedback. Special thanks to Halley for always talking about sci-fi movies with me.

Finally, a huge thank you to New Degree Press, especially Eric Koester, Brian Bies, Hazel L Lorico, and Anastasia Armendariz, for helping me write a book on topics I am extremely passionate about.

APPENDIX

INTRODUCTION

Lardinois, Frederic. 2018. "Google Says It Sold A Google Home Device Every Second Since October 19". *Techcrunch.* https://techcrunch.com/2018/01/05/google-says-it-sold-a-google-home-device-every-second-since-october-19/.

"Pure Imagination [From Willy Wonka And The Chocolate Factory] Lyrics". 2019. *Lyrics.Com.* Accessed January 15. https://www.lyrics.com/lyric/27628631.

Reisinger, Don. 2017. "Http://Fortune.Com". *Fortune.* http://fortune.com/2017/03/06/apple-iphone-use-worldwide/.

PART 1

"The Avengers (2012)". 2019. *Imdb*. Accessed January 13. https://www.imdb.com/title/tt0848228/characters/nm0000569.

Lincoln, Don. 2017. "Unfortunately, The Force May Not Be With You". *CNN*. https://www.cnn.com/2017/12/17/opinions/star-wars-science-possibilities-opinion-lincoln/index.html.

Lincoln, Don. 2015. "Is A Real Lightsaber Possible? Science Offers A New Hope". *Space.Com*. https://www.space.com/31361-building-a-real-lightsaber.html.

"Quotes - Help Me, Obi-Wan Kenobi. You're My Only Hope.". 2019. *Shmoop.Com*. Accessed January 13. https://www.shmoop.com/quotes/help-me-obi-wan-kenobi-only-hope.html.

"Scotty, I Need Warp Speed In Three Minutes Or We're All Dead! - Star Trek: The Wrath Of Khan". 2019. *Quotegeek*. Accessed January 15. http://quotegeek.com/quotes-from-movies/star-trek-the-wrath-of-khan/840/.

"Star Wars: Solo's Most Important Quote And What It Means". 2019. *Den Of Geek*. Accessed January 13. https://www.denofgeek.com/us/movies/star-wars/273785/star-wars-solos-most-important-quote-and-what-it-means.

"View Quote ... Star Wars Episode IV: A New Hope ... Movie Quotes Database". 2019. *Moviequotedb.Com*. Accessed January 13. http://www.moviequotedb.com/movies/star-wars-episode-iv-a-new-hope/quote_29884.html.

"View Quote ... Star Wars Episode V: The Empire Strikes Back ... Movie Quotes Database". 2019. *Moviequotedb. Com*. Accessed January 13. http://www.moviequotedb.com/movies/star-wars-episode-v-the-empire-strikes-back/quote_22681.html.

PART 2

"Guardians Of The Galaxy Vol. 2 - Wikiquote". 2019. *En.Wikiquote.Org*. Accessed January 7. https://en.wikiquote.org/wiki/Guardians_of_the_Galaxy_Vol._2.

"Sleep In Adolescents". 2019. *Nationwidechildrens.Org*. Accessed January 3. https://www.nationwidechildrens.org/specialties/sleep-disorder-center/sleep-in-adolescents.

"Star Wars: Episode III - Revenge Of The Sith - Movie Quotes - Rotten Tomatoes". 2019. *Rottentomatoes.Com*. Accessed January 7. https://www.rottentomatoes.com/m/star_wars_episode_iii_revenge_of_the_sith/quotes/.

"View Quote ... Star Wars Episode V: The Empire Strikes Back ... Movie Quotes Database". 2019. *Moviequotedb. Com*. Accessed January 7. http://www.moviequotedb. com/movies/star-wars-episode-v-the-empire-strikes-back/ quote_22697.html.

"Yondu Udonta/Quote". 2019. *Marvel Cinematic Universe Wiki*. Accessed January 7. http://marvelcinematicuniverse. wikia.com/wiki/Yondu_Udonta/Quote.

PART 3

"Big Hero 6 - Calgary Public Library". 2019. *Bibliocommons*. Accessed January 7. https://calgary.bibliocommons.com/item/ quotation/987362095.

"Cyberbullying Facts And Statistics - Teensafe". 2016. *Teensafe*. https://www.teensafe.com/blog/ cyber-bullying-facts-and-statistics/.

"Excuse Me Sir, But That R2-D2 Is In Prime Condition, A Real Bargain.". 2019. *Imgur*. Accessed January 7. https://imgur. com/gallery/MAu9joS.

"From Here To Eternity (1953)". 2019. *Imdb*. Accessed January 7. https://www.imdb.com/title/tt0045793/characters/ nm0001050.

Gallup, Inc. 2017. "The World's Broken Workplace". *Gallup. Com.* https://news.gallup.com/opinion/chairman/212045/world-broken-workplace.aspx?g_source=position1&g_medium=related&g_campaign=tiles.

Gherini, Annie. 2018. "Gen-Z Is About To Outnumber Millennials. Here's How That Will Affect The Business World". *Inc.Com.* https://www.inc.com/anne-gherini/gen-z-is-about-to-outnumber-millennials-heres-how-that-will-affect-business-world.html.

Green, Yasmin. 2018. *How Technology Can Fight Extremism And Online Harassment.* Video. TED.

Gruber, Tom. 2017. *How AI Can Enhance Our Memory, Work And Social Lives.* Video. TED.

"J.A.R.V.I.S./Quote". 2019. *Marvel Cinematic Universe Wiki.* Accessed January 7. http://marvelcinematicuniverse.wikia.com/wiki/J.A.R.V.I.S./Quote.

"The Matrix: Quotes About Versions Of Reality Page 2". 2019. *Shmoop.Com.* Accessed January 7. https://www.shmoop.com/the-matrix/versions-of-reality-quotes-2.html.

Lee, David. 2017. *Why Jobs Of The Future Won't Feel Like Work.* DVD. TED@UPS.

Li, Fei Fei. 2015. *How We're Teaching Computers To Understand Pictures*. DVD. TED.

Ratcliffe, Amy. 2015. "6 Of C-3PO's Best Insults | Starwars. Com". *Starwars.Com*. Accessed January 7. https://www. starwars.com/news/6-of-c-3pos-best-insults.

"Robin Williams' Best Dead Poets Society Quotes: 'Carpe Diem. Seize The". 2019. *The Independent*. Accessed January 7. https://www.independent.co.uk/arts-entertainment/films/ news/robin-williams-best-dead-poets-society-quotes-carpe- hear-it-carpe-carpe-diem-seize-the-day-boys-9663800.html.

Sullivan, Robert. 2017. "Meet The Head International Troll Slayer At Google". *Vogue*. https://www.vogue.com/article/ google-jigsaw-yasmin-green-internet-trolls-web-security.

"View Quote ... Star Wars Episode IV: A New Hope ... Movie Quotes Database". 2019. *Moviequotedb.Com*. Accessed January 13. http://www.moviequotedb.com/movies/star-wars- episode-iv-a-new-hope/quote_29884.html.

"View Quote ... Star Wars Episode V: The Empire Strikes Back ... Movie Quotes Database". 2019. *Moviequotedb. Com*. Accessed January 7. http://www.moviequotedb. com/movies/star-wars-episode-v-the-empire-strikes-back/ quote_22697.html.

"Vision/Quote". 2019. *Marvel Cinematic Universe Wiki.* Accessed January 7. http://marvelcinematicuniverse.wikia. com/wiki/Vision/Quote.

PART 4

"J.A.R.V.I.S./Quote". 2019. *Marvel Cinematic Universe Wiki.* Accessed January 7. http://marvelcinematicuniverse.wikia. com/wiki/J.A.R.V.I.S./Quote.

Sachdev, Umesh. 2017. *The Future Of Voice Technology.* Video. TED.

"Winter Soldier/Quote". 2019. *Marvel Cinematic Universe Wiki.* Accessed January 7. http://marvelcinematicuniverse. wikia.com/wiki/Winter_Soldier/Quote.

PART 5

"A Quote From A Game Of Thrones". 2019. *Goodreads. Com.* Accessed January 7. https://www.goodreads.com/ quotes/227324-a-mind-needs-books-as-a-sword-needs-a.

"The Best Samwell Tarly Quotes". 2019. *Ranker.* Accessed January 7. https://www.ranker.com/list/best-samwell-tarly-quotes/ ranker-of-thrones.

"Doctor Strange (2016)". 2019. *Imdb*. Accessed January 7. https://www.imdb.com/title/tt1211837/characters/nm0842770.

"Heimdall/Quote". 2019. *Marvel Cinematic Universe Wiki*. Accessed January 7. http://marvelcinematicuniverse.wikia. com/wiki/Heimdall/Quote.

McAteer, John. 2019. ""We Never Lose Our Demons, We Only Learn To Live Above Them."". *Video Ut Intellectum*. Accessed January 7. https://filmphilosopher.wordpress. com/2016/11/06/we-never-lose-our-demons-we-only-learn-to-live-above-them/.